Safer Healthcare

Charles Vincent · René Amalberti

Safer Healthcare

Strategies for the Real World

 Springer Open

Charles Vincent
University of Oxford
Oxford, United Kingdom

René Amalberti
Haute Autorité de Santé
Paris, France

This work is supported by the Health Foundation, an independent charity working to improve the quality of healthcare.

ISBN 978-3-319-25557-6   ISBN 978-3-319-25559-0   (eBook)
DOI 10.1007/978-3-319-25559-0

Library of Congress Control Number: 2015957594

Springer Cham Heidelberg New York Dordrecht London

Printed on acid-free paper

Springer International Publishing AG Switzerland is part of Springer Science+Business Media (www.springer.com)

To Lucian Leape and James Reason

Preface

Healthcare has brought us extraordinary benefits, but every encounter and every treatment also carries risk of various kinds. The known risks from specific treatments are well established and routinely discussed by clinicians. Yet we also face risks from failures in the healthcare system, some specific to each setting and others from poor coordination of care across settings. For us, as patients, healthcare provides an extraordinary mixture of wonderful achievements and humanity which may be rapidly followed by serious lapses and adverse effects.

Patient safety has been driven by studies of specific incidents in which people have been harmed by healthcare. Eliminating these distressing, sometimes tragic, events remains a priority, but this ambition does not really capture the challenges before us. While patient safety has brought many advances, we believe that we will have to conceptualise the enterprise differently if we are to advance further. We argue that we need to see safety through the patient's eyes, to consider how safety is managed in different contexts and to develop a wider strategic and practical vision in which patient safety is recast as the management of risk over time.

The title may seem curious. Why 'strategies for the real world'? The reason is that as we developed these ideas we came to realise that almost all current safety initiatives are either attempts to improve the reliability of clinical processes or wider system improvement initiatives. We refer to these as 'optimising strategies', and they are important and valuable initiatives. The only problem is that, for a host of reasons, it is often impossible to provide optimal care. We have very few safety strategies which are aimed at managing risk in the often complex and adverse daily working conditions of healthcare. The current strategies work well in a reasonably controlled environment, but they are in a sense idealistic. We argue in this book that they need to be complemented by strategies that are explicitly aimed at managing risk 'in the real world'.

How the Book Came to Be Written

We are friends who have been passionate about safety for many years. We did not meet however until we were invited as faculty members to the memorable Salzburg International seminar on patient safety organised by Don Berwick and Lucian Leape in 2001.

The story of the book began in late 2013 with René's observation that the huge technological and organisational changes emerging in healthcare would have considerable implications for patient safety. Charles suggested that care provided in the home and community were an important focus and we planned papers addressing these subjects. We began to speak and meet on a regular basis, evolving a common vision and set of ideas in numerous emails, telephone calls and meetings. It quickly became evident that a new vision of patient safety was needed now, and that the emerging changes would just accelerate the present requirements. We needed a book to express these ideas in their entirety.

The particular characteristic of this book is that it has been really written by 'four hands'. In many jointly written books, chapters have clearly been divided between authors. In contrast, we made no specific allocations of chapters to either of us at any point. All chapters were imagined and developed together, and the ideas tested and hammered into shape by means of successive iterations and many discussions.

The work matured slowly. The essential ideas emerged quite quickly but it was challenging to find a clear expression, and the implications were much broader than we had imagined. We were also determined to keep the book short and accessible and, as is widely recognised, it is much harder to write a short book than a long one. We completed a first draft in April 2015 which was read by generous colleagues and presented to an invited seminar at the Health Foundation. We received encouragement and enthusiasm and much constructive comment and criticism which helped us enormously in shaping and refining the final version which was delivered to Springer in August 2015.

The Structure of the Book

In the first chapter of this book, we set out some of the principal challenges we face in improving the safety of healthcare. In the second, we outline a simple framework describing different standards of healthcare, not to categorise organisations as good or poor, but suggesting a more dynamic picture in which care can move rapidly from one level to another. We then argue that safety is not, and should not, be approached in the same way in all clinical environments; the strategies for managing safety in highly standardised and controlled environments are necessarily different from those in which clinicians must constantly adapt and respond to changing circumstances. We then propose that patient safety needs to be seen and understood from the perspective of the patient. We are not taking this perspective in order to respond to policy imperatives or demands for customer focus but simply because that is the reality we need to capture. Safety from this perspective involves mapping the risks and benefits of care along the patient's journey through the healthcare system.

The following chapters begin to examine the implications of these ideas for patient safety and the management of risk. In Chap. 5, we build on our previous understanding of the analysis of incidents to propose and illustrate how analyses across clinical contexts and over time might be conducted. The role of the patient and family in selection, analysis and recommendations is highlighted.

Chapter 6 outlines an architecture of safety strategies and associated interventions that can be used both to manage safety on a day-to-day basis and to improve safety over the long term. The strategies are, we believe, applicable at all levels of the healthcare system from the frontline to regulation and governance of the system. As we have mentioned, most safety improvement strategies aim to optimise care. Within this general approach, we distinguish focal safety programmes aimed at specific harms or specific clinical processes and more general attempts to improve work systems and processes. We suggest that these strategies need be complemented by strategies that are more concerned with detecting and responding to risk and which assume, particularly in a time of rising demand and financial austerity, that care will often be delivered in difficult working conditions. These three additional approaches are: risk control; monitoring, adaptation and response; and mitigation. Clinicians, managers and others take action every day to manage risk but curiously this is not generally seen as patient safety. We need to find a vision that brings all the potential ways of managing risk and safety into one broad frame. Optimisation strategies improve efficiency and other aspects of quality as much as they improve safety. In contrast, risk control, adaptation and recovery strategies are most concerned with improving safety.

In Chaps. 7, 8, and 9 we explore the use and value of this strategic framework and consider how safety should be addressed in hospitals, home and in primary care, paying particular attention to safety in the home. We have found it difficult to make hospitals safe, even with a highly trained and professional workforce within a relatively strong regulatory framework. We will shortly be trying to achieve similar standards of safety with a largely untrained workforce (patients and their carers) in settings not designed for healthcare and with almost no effective oversight or supervision. This may prove challenging.

We believe that an expanded vision of patient safety is needed now. However in Chap. 10 we argue that the forthcoming changes in the nature, delivery and organisational forms of healthcare make the transition even more urgent. The healthcare of the future, with much more care being delivered in the home under the patient's direct control, will require a new vision of patient safety necessarily focused on patients and their environment more than on professionals and the hospital environment. Discussions of new technologies and the potential for care being delivered in a patient's home are generally marked by unbridled optimism without any consideration of new risks that will emerge or the potential burden on patients, family and carers as they take on increasing responsibilities. The new scenario will bring great benefits, but also new risks which will be particularly prominent during the transitional period. For an active patient with a single chronic illness, empowerment and control of one's treatment may be an unalloyed benefit, provided professionals are available when required. When one is older, frail or vulnerable, the calculation of risk and benefit may look very different.

In the final two chapters, we draw all the material together and present a compendium of all the safety strategies and interventions discussed in this book. We describe this as an 'incomplete taxonomy' as we are conscious that, if this approach is accepted, there is much to be done to map the landscape of strategies and interventions. These interventions can be selected, combined and customised to context. We hope that this framework will support frontline leaders, organisations, regulators and government in devising an effective overall strategy for managing safety in the face of austerity and rising demand. In the final chapter, we set out some immediate directions and implications for patients, clinical staff and managers, executives and boards, and those concerned with regulation and policy. Financial pressures and rising demand can often distract organisations from safety and quality improvement which can temporarily become secondary issues. In contrast, we believe that financial pressures provoke new crises in safety and that we urgently need an integrated approach to the management of risk.

We know that these ideas need to be tested in practice and that ultimately the test is whether this approach will lead in a useful direction for patients. We believe very strongly that the proposals we are making can only become effective if a community of people join together to develop the ideas and implications.

Oxford, UK   Charles Vincent
Paris, France   René Amalberti

Acknowledgements and Thanks

We gained great encouragement from the initial responses to an earlier draft of the book. We also received a host of ideas, suggestions and insightful comments which illuminated specific issues or identified flaws, infelicities and things that were just plain wrong. Where we have included specific quotes or examples provided by individuals we have cited them in the text, but all the comments we received were valuable and led to important changes to both the structure and the content of the book. The book you see today is very different from the draft originally circulated. We would like to thank the following people for their insights, suggestions and constructive criticism: Jill Bailey, Nick Barber, Maureen Bisognano, Jane Carthey, Bryony Dean Franklin, John Green, Frances Healey, Goran Henriks, Ammara Hughes, Matt Inada Kim, Jean Luc Harousseau, John Illingworth, Martin Marshall, Phillipe Michel, Wendy Nicklin, Penny Pereira, Anthony Staines and Suzette Woodward.

The Health Foundation is remarkable in encouraging the development of new ideas and giving people the freedom and time to attack challenging problems. We thank the Health Foundation for their enthusiasm and support for this book. Charles would in particular like to thank Jennifer Dixon, Nick Barber, Jo Bibby, Helen Crisp and Penny Pereira for enabling a career transition and for their personal support and encouragement over many years. Michael Howes brought life and colour to our tentative figures. We thank Nathalie Huilleret at Springer for her enthusiasm for the project, her personal oversight of publication and her willingness and encouragement to make this book Open Access and available to all.

Charles Vincent
René Amalberti
Oxford and Paris
August 2015

Contents

List of Figures

Progress and Challenges for Patient Safety

Twenty-five years ago the field of patient safety, apart from a number of early pioneers, did not exist and the lack of research and attention to medical accidents could reasonably be described as negligent (Vincent 1989). There is now widespread acceptance and awareness of the problem of medical harm and, in the last decade, considerable efforts have been made to improve the safety of healthcare. Progress has however been slower than originally anticipated and the earlier optimism has been replaced by a more realistic longer-term perspective. There has undoubtedly been substantial progress but we believe that future progress, particularly in the wider healthcare system, will require a broader vision of patient safety. In this chapter we briefly review progress on patient safety and consider the principal future challenges as we see them.

Progress on Patient Safety

With the massive attention now given to patient safety it is easy to forget how difficult it was in earlier years to even find clear accounts of patient harm, never mind describe and analyse them. Medico-legal files, oriented to blame and compensation rather than safety, were the principal source of information (Lee and Domino 2002). In contrast narrative case histories and accompanying analyses and commentary are now widely available. Analyses of incidents are now routinely performed, albeit often in a framework of accountability rather than in the spirit of reflection and learning.

Major progress has been made in assessing the nature and scale of harm in many countries. The findings of the major record review studies are widely accepted (de Vries et al. 2008) and numerous other studies have catalogued the nature and extent of surgical adverse events, infection, adverse drug events and other safety issues. The measurement and monitoring of safety continues to be a challenge but progress has been made in developing reliable indicators of safety status (Vincent et al. 2013, 2014).

© The Author(s) 2016
C. Vincent, R. Amalberti, *Safer Healthcare: Strategies for the Real World*,
DOI 10.1007/978-3-319-25559-0_1

Analyses of safety incidents have revealed a wide range of contributory factors and that individual staff are often the inheritors of wider system problems (Reason 1997). However, some safety problems can be linked to the sub-standard performance of individuals, whether wilful or due to sickness or incapacity (Francis 2012). Regulation of both organisations and individuals is increasing and every healthcare professional now has a clear duty to report consistent poor performance from a colleague. Drawing attention to safety issues is actively encouraged at the highest levels, although many whistle-blowers are still shabbily treated and persecuted for their efforts. All of these developments represent an increasing concern with safety and determination to improve basic standards.

Substantial progress has also been made in mapping and understanding safety issues. Surgery, for instance, was long ago identified as the source of a high proportion of preventable adverse events. A decade ago most of these would have been considered unavoidable or ascribed, generally incorrectly, as due to poor individual practice (Calland et al. 2002; Vincent et al. 2004). Studies of process failures, communication, teamwork, interruptions and distractions have now identified multiple vulnerabilities in surgical care. Given the inherent unreliability of the system it now seems remarkable that there are so few adverse events, which is probably testament to the resilience and powers of recovery of clinical staff (Wears et al. 2015). Many surgical units are now moving beyond the undoubted gains of checklists to consider the wider surgical system and the need for a more sophisticated understanding of surgical teamwork in both the operating theatre and the wider healthcare system (de Vries et al. 2010).

A considerable number of interventions of different kinds have shown that errors can be reduced and processes made more reliable. Interventions such as computer order entry, standardisation and simplification of processes and systematic handover have all been shown to improve reliability, and in some cases reduce harm, in specific contexts. We have however relatively few examples of large scale interventions which have made a demonstrable impact on patient safety, the two most notable exceptions being the reduction of central line infections in Michigan and the introduction of the WHO surgical safety checklist (Pronovost et al. 2006; Haynes et al. 2009) (Table 1.1).

While specific interventions have been shown to be effective it has proved much more difficult to improve safety across organisations. The United Kingdom Safer Patients Initiative, which engaged some of the acknowledged leaders in the field, was one of the largest and most carefully studied intervention programmes. The programme was successful in many respects, in that it engaged and energised staff and produced pockets of sustained improvement. However it failed to demonstrate large scale change on a variety of measures of culture, process and outcomes (Benning et al. 2011). Similarly, where studies have attempted to assess safety across a whole healthcare system, the findings have generally been disappointing. Longitudinal record review studies in the United States, France have shown no improvement in patient safety although there have recently been encouraging results from Netherlands (Landrigan et al. 2010; Michel et al. 2011; Baines et al. 2015)

Table 1.1 Progress in patient safety over two decades

	Where we were (1995)	Where we are now (2015)
Foundations	Incident reporting, continuous improvement and development of best practice	Largely unchanged. More translation and use of industrial approaches to safety, increased attention to incident analysis, learning and feedback
Definition	Harm defined from a professional standpoint, rooted in a medico-legal and insurance perspective. Narrow vision of causality, direct association between technical care and harm	Patient safety still linked to a medico-legal perspective. Broader understanding of human error and organisational influences
Perimeter of inclusion	Dominant technical vision of care, improved clinical protocols as main priority for improving safety	Recognition of the importance of human factors and human sciences. Organisational factors and safety culture are additional priorities for safety
Measurement	Counting incidents and adverse events	Largely unchanged

Compared to a decade ago, we now have a good understanding of the phenomenology of error and harm, a considerable amount of epidemiological data, some understanding of the causes of harm, demonstrations of the efficacy of certain interventions and the effectiveness of a few. We do not have clear evidence of wide sustained change or widespread improvements in the safety of healthcare systems. All in all, progress looks reasonable if not spectacular. Given the level of resources allocated to safety, still tiny in comparison with biomedicine, progress looks reasonably good.

We believe that the concept of patient safety we are working with is too narrow and that future progress, particularly outside hospitals, will require a broader vision. In the remainder of this chapter we set out some challenges and confusions that we regard as particularly critical. These provide both the motivation for our work together and also an introduction to our approach.

Harm Has Been Defined Too Narrowly

We agree with those who seek to provide a more positive vision of safety (Hollnagel 2014). The punitive approach sometimes taken by governments, regulators and the media is, for the most part, deeply unfair and damaging. Healthcare while enormously beneficial is, like many other important industries, also inherently hazardous. Treating patients safely as well as effectively should be regarded as an achievement and celebrated.

We make no apologies however for continuing to focus on harm as the touchstone for patient safety and the motivation for our work. We will put up with errors and problems in our care, to some extent at least, as long as we do not come to harm and the overall benefits clearly outweigh any problems we may encounter. Many errors do not lead to harm and may even be necessary to the learning and maintenance of safety. Surgeons, for example, may make several minor errors during a procedure, none of which really compromise the patient's safety or the final outcome of the operation.

Patient safety, particularly the large scale studies of adverse events, has its origins in a medico-legal concept of harm. We have, for the most part, now separated the concept of harm from that of negligence which is an important achievement, though we still tend to think of safety as being the absence of specific harmful or potential harmful events (Runciman et al. 2009). Harm can also result from loss of opportunity due to a combination of poor care and poor coordination whether inside the hospital, at the transition with primary care, or over a long period of time in the community. Evidence is growing that many patients suffer harm, in the sense that their disease progresses untreated, through diagnostic error and delay (Graber 2013; Singh et al. 2014). In some contexts, this would simply be seen as poor quality care falling below the accepted standard. But for the patient a serious failure can lead to untreated or unrecognised disease and, from their perspective, to harm.

Box 1.1 Safety Words and Concepts

The term 'medical error' has been used in a variety of ways, often as shorthand for a poor outcome. We use the term error is in its everyday sense as a retrospective judgement that an action or omission by a person did not achieve the intended outcome. We use the term reliability when considering processes and systems rather than the actions of people.

The aims of the patient safety movement can be stated in a number of different ways:

- To reduce harm to patients, both physical and psychological
- To eliminate preventable harm
- To reduce medical error
- To improve reliability
- To achieve a safe system

All these are reasonable objectives but they are subtly different. We suggest that the central aim must be to prevent or at least reduce harm to patients, while acknowledging that the concept of harm is difficult to define and other objectives are also valid. As the book develops we will suggest that the most productive way to approach patient safety is to view it as the management of risk over time in order to maximise benefit and minimise harm to patients in the healthcare system.

We believe that the current focus on specific incidents and events is too narrow and that we need to think about harm much more broadly and within the overall context of the benefits of treatment. As the book evolves, we endeavour to develop a different vision which is more rooted in the experience of patients. As patients, the critical question for us is to weigh up the potential benefits against the potential harms which may, or may not, be preventable. While we certainly want to avoid harmful incidents, we are ultimately concerned with the longer term balance of benefit and harm that accrues over months or years or even over a lifetime.

Safety Is a Moving Target

Safety is, in a number of respects, a constantly moving target. As standards improve and concern for safety grows within a system, a larger number of events will come to be considered as safety issues. In a very real sense innovation and improving standards create new forms of harm in that there are new ways the healthcare system can fail patients.

In the 1950s many complications of healthcare were recognised, at least by some, but largely viewed as the inevitable consequences of medical intervention (Sharpe and Faden 1998). Gradually, certain types of incidents have come to seem both unacceptable and potentially preventable. The clearest example in recent times is healthcare-associated infection, which is no longer viewed as an unfortunate side effect of healthcare. With increased understanding of underlying processes, mechanisms of transmission and methods of prevention, coupled with major public and regulatory pressure, such infections are becoming unacceptable to both patients and professionals (Vincent and Amalberti 2015).

In the last 10 years, as more types of harm have come to be regarded as preventable, the perimeter of patient safety has expanded. A larger number of harmful events are now regarded as 'unacceptable'. In addition to infections we could now include, in the British NHS, pressure ulcers, falls, venous thromboembolism and catheter associated urinary tract infections. In the United Kingdom the Francis Report into Mid Staffordshire Hospitals NHS Trust highlighted additional risks to patients, such as malnutrition, dehydration and delirium all of which are now being viewed as safety issues. We should also consider adverse drug reactions in the community that cause admission to hospital, polypharmacy and general harm from overtreatment. All these, in the past, might have been regretted but are now receiving greater attention through being viewed under the safety umbrella.

The perimeter of safety is therefore expanding but this does not mean that healthcare is becoming less safe. A long-standing concern with safety in such specialties as anaesthesia and obstetrics is actually a marker of the high standards these specialties have achieved. Safety is an aspiration to better care and labelling an issue as a safety issue is a strongly motivational, sometimes emotional, plea that such outcomes cannot and should not be tolerated (Vincent and Amalberti 2015).

Only Part of the Healthcare System Has Been Addressed

Patient safety has evolved and developed in the context of hospital care. The understanding we have of the epidemiology of error and harm, the causes and contributory factors and the potential solutions are almost entirely hospital based. The concepts which guided the study of safety in hospitals remain relevant in primary and community care but new taxonomies and new approaches may be required in these more distributed forms of healthcare delivery (Brami and Amalberti 2010; Amalberti and Brami 2012).

Care provided in a person's home is an important context for healthcare delivery but patient safety in the home has not been addressed in a systematic manner. The home environment may pose substantial risks to patients, greater in some cases than in the hospital environment. Safety in the context of a patient's home care is likely to require different concepts, approaches and solutions to those developed in the hospital setting. This is because of the different environment, roles, responsibilities, standards, supervision and regulatory context in home care. Critical differences are that patients and carers are autonomous and are increasingly taking on professional roles; they rather than the professional become the potential source of medical error. Additionally, stressful and potentially hazardous conditions, such as poor lighting, mean that socio-economic conditions take on a much greater importance.

In both primary care and care at home the risks to patients are rather different from those in hospital, being much more concerned with omissions of care, failure to monitor over long time periods and lack of access to care. These areas have not traditionally fallen within the area of patient safety but are undoubtedly sources of potential harm to patients. The concept of the patient safety incident, and even of adverse events, breaks down in these settings or is at least stretched to its limit. Suppose, to take just one example, a patient is hospitalised after taking an incorrect dose of warfarin for 4 months. The admission to hospital could be viewed as an incident or a preventable adverse event. This description however hardly does justice to 4 months of increasing debility and ill health culminating in a hospital admission. In reality, the admission to hospital is the beginning of the recovery process and a sign that the healthcare system is at last meeting the needs of this patient. The episode needs to be seen not as an isolated incident but as an evolving and prolonged failure in the care provided to this person.

We Are Approaching Safety in the Same Way in All Settings

'But we are not like aviation' someone will inevitably say in any discussion of the value of learning from commercial aviation and comparing approaches to safety in different sectors. Well no, healthcare is not like aviation in any simple sense. But some aspects of healthcare are comparable to some aspects of aviation. An surgical operation does not have a great deal in common with a commercial flight but the pre-flight checking process is comparable to the pre-operation checking process and so learning how aviation manages those checks is instructive.

The objection to the simple comparison is important. Safety in healthcare does need to be approached differently from safety in commercial aviation. The wholesale transfer of aviation approaches to healthcare at the very least requires considerable adaptation; crew resource management acted as an inspiration to surgical and anaesthetics teams but surgical team training has now developed its own style and history (Gaba 2000; Sevdalis et al. 2009). We now need to go further and consider a still more important issue which is that safety may need to be approached differently in different areas of healthcare. Specialty specific approaches (Croskerry et al. 2009) are emerging but models, methods and interventions do not often distinguish between settings.

Healthcare is a particularly complex environment. We might say that healthcare is 20 different industries under one banner. Consider the hospital environment with multiple types of work, many different professions and varying working conditions across clinical environments. There are areas of highly standardized care such as pharmacy, radiotherapy, nuclear medicine and much of the process of blood transfusion. All of these are highly standardized and rely heavily on automation and information technology. They are islands of reliability within the much more chaotic wider hospital environment. On the ward standards and protocols provide important controls on hazards (such as infection from poor hand hygiene) but day-to-day conditions demand constant adaptation and flexibility. Other sections of the hospital, such as the emergency department, continually have to deal with unpredictable patients flow and workloads; their activity needs considerable hour-by-hour adaptation because of the huge variety of patients, the complexity of their conditions and the vulnerabilities of the healthcare system.

The risks and the nature of the work vary across all these settings. In spite of this we are essentially using the same concepts, the same analytic toolbox and the same suite of interventions in all settings. Many of these approaches can be customised and adapted to different settings. However we will argue later in the book that risk needs to be managed in very different ways in different environments and that the approach of, for instance, commercial aviation is very different from that of professionals working in more fluid risky environments such as fire-fighters. In healthcare we may have to adapt our approach to safety according to the nature of the work, the working conditions and use a variety of underlying models of safety.

Our Model of Intervention Is Limited

The most dramatic safety improvements so far demonstrated have been those with a strong focus on a core clinical issue and a relatively narrow timescale. These interventions, such as the surgical safety checklist and the control of central line infections, are of course far from simple in the sense that they have only succeeded because of a sophisticated approach to clinical engagement and implementation. More general system improvements may extend to an entire patient pathway. For instance the introduction of the SURPASS system using checklists and other improvements to communication along the entire surgical pathway and showed a reduction in surgery complications (de Vries et al. 2010). Bar coding and other systems have massively enhanced the reliability of blood transfusion systems, incrementally improving each step of the pathway (Murphy et al. 2013).

We should however be wary of modelling all future safety interventions on our most visible successes. At the moment the primary focus is on developing interventions to address specific harms or to improve reliability at specific points in a care process. This, entirely reasonable, approach is evolving to include the reliability of entire care pathways or areas of care (such as an out-patient clinic). We will argue however that, in addition to increasing reliability, we also need to develop proactive strategies to manage risk on an ongoing basis, particularly in less controlled environments. There is also a class of strategies and interventions, particularly those that focus on detecting and responding to deviations, that are particularly critical for preventing harm to patients. These approaches do not feature as strongly in the classical quality and safety armament.

We also need to recognise that safety, for any person or organisation, is always only one of a number of objectives. For instance, many sports involve an element of risk and potential harm. When we become patients we necessarily accept the risks of healthcare in pursuit of other benefits. Similarly a healthcare organisation can never treat safety as the sole objective, even if they say safety is their 'top priority'. Of necessity, safety is always only one consideration in a broader endeavour, whether in healthcare or in any other field. As an oil executive expressed it: 'Safety is not our top priority. Getting oil out of the ground is our priority. However, when safety and productivity conflict, then safety takes precedence' (Vincent 2010). Similarly, in healthcare, the main objective is providing healthcare to large numbers of people at a reasonable cost, but this needs to be done as safely as possible.

Healthcare Is Changing

We have argued that, for a variety of reasons, we need to expand our view of patient safety. This argument has been made from our understanding of current healthcare systems. However we also believe that the rapid evolution of healthcare, combined with increasing financial pressures, brings an additional urgency to the quest for a new vision.

Outcomes of care have improved rapidly all over the world. People now survive illnesses, such as myocardial infarction and stroke, which were once fatal. As the effectiveness of healthcare improves, increasing numbers of patients are ageing with their illness under control. Current projections suggest that by 2030 as many as 25 % of the population in many countries may surviving into their 90s. In many cases an illness which was once fatal has become a chronic condition with all the related implications for the individual, society and the healthcare system. The treatment of chronic conditions (such as diabetes, respiratory diseases, depression, cardiac and renal disease) is now the major priority. The phenomenal increase in diabetes alone (although not driven by ageing per se) threatens to destabilise healthcare systems and the general increase in multiple comorbidities and more complex health problems places huge stress on healthcare systems. The question of what 'best practice' actually is for any individual patient is itself becoming a very difficult issue.

The impact on global cost of healthcare is considerable, with average costs increasing by 1 % of national gross domestic product (GDP) between 2000 and 2013 (World Bank). By 2030 there may be 30 % more patients with chronic conditions which might require a further increase in funding of between 2 and 4 % of GDP, depending on the approach taken by the country in question. There is a major risk that by 2030, institutional care for the aged will be unaffordable and that, in the absence of alternatives, there will be a crisis of quality in care for the aged. While alternative systems are evolving there could be if anything an increased risk of failures and harm to patients.

The need for healthcare to evolve and adapt is to a very large extent the consequence of the successes of modern medicine. The focus of care needs to move rapidly from high quality care in hospitals to a focus on the entire patient journey over years or even over a lifetime. These changes are long overdue but becoming increasingly urgent. The shift to the management of care over long time periods and many settings has a number of consequences with implications for safety. Patients stay in hospital less time, live at home for years with their disease, with a consequent transfer of responsibility from hospitals to primary care. This requires effective coordination across all health care organisations, in particular at the transition points, in order to mitigate risk and enable positive outcomes. Reducing complexity is crucial.

Finally, patients are more knowledgeable and informed than previously. They are increasingly aware of their rights to information and access. The public expects a system that meets their needs in a holistic and integrated way, with a seamless system of effective communication between transition points. Last but not least there is an increasing emphasis on the prevention of disease and the maintenance of health. This turns the concept of the patient journey into the concept of the citizen or person journey.

The combination of austerity, rising healthcare costs, rising standards and increased demand will place huge pressures on healthcare systems which will increase the likelihood of serious breakdowns in care. At the same time innovations in the delivery of care in the home and community, while providing new benefits, will also create new forms of risk. Our current models of safety are not well adapted to this new landscape.

In this chapter we identified a number of challenges for patient safety. In the next three chapters we begin to consider how these challenges are to be met and establish the foundations for the more practical and strategic chapters that follow later in the book. First however we build the foundations beginning with the simple idea that care given to patients is of varying standard and, equally important, that the care given to any one patient varies considerably along their journey.

Key Points

- Major progress has been made in assessing the nature and scale of harm to patients in many countries
- A considerable number of interventions of different kinds have shown that errors can be reduced and processes made more reliable.
- The most safety improvements so far demonstrated have been those with a strong focus on a core clinical issue and a relatively narrow timescale. It has proved very much more difficult to improve safety across whole organisations
- Improving safety at a population level has been even more challenging and findings have generally been disappointing.
- Safety is, in a number of respects, a constantly moving target. The perimeter has expanded over time as new forms of harm have been identified as safety issues.
- Patient safety has evolved and developed in the context of hospital care. The concepts which guided the study of safety in hospitals remain relevant in primary and community care but new approaches to safety will be required in these more distributed forms of healthcare delivery
- The successes of healthcare and improved living conditions mean that people live longer with chronic conditions which were once fatal. This has led to considerable transfer of responsibility from hospitals to home and primary care. Safety models, safety methods, and interventions strategies must change accordingly.
- The combination of austerity, rising healthcare costs, rising standards and increased demand will place huge pressures on healthcare systems which will increase the likelihood of serious breakdowns in care. Innovations in the delivery of care in the home and community, while providing new benefits, will also create new forms of risk. Our current models of safety are not well adapted to this new landscape.

References

Amalberti R, Brami J (2012) 'Tempos' management in primary care: a key factor for classifying adverse events, and improving quality and safety. BMJ Qual Saf 21(9):729–736

Baines R, Langelaan M, de Bruijne M, Spreeuwenberg P, Wagner C (2015) How effective are patient safety initiatives? A retrospective patient record review study of changes to patient safety over time. BMJ Qual Saf. doi:10.1136/bmjqs-2014-003702

Benning A, Dixon-Woods M, Nwulu U, Ghaleb M, Dawson J, Barber N, Franklin BD, Girling A, Hemming K, Carmalt M, Rudge G, Naicker T, Kotecha A, Derrington MC, Lilford R (2011)

Multiple component patient safety intervention in English hospitals: controlled evaluation of second phase. BMJ 342:d199. doi:10.1136/bmj.d199

Brami J, Amalberti R (2010) La sécurité du patient en médecine générale (Patient safety in primary care). Springer Science & Business Media, Paris

Calland JF, Guerlain S, Adams RB, Tribble CG, Foley E, Chekan EG (2002) A systems approach to surgical safety. Surg Endosc Other Interv Tech 16(6):1005–1014

Croskerry P, Cosby KS, Schenkel SM, Wears RL (eds) (2009) Patient safety in emergency medicine. Lippincott Williams & Wilkins, Philadelphia

de Vries EN, Prins HA, Crolla RM, den Outer AJ, van Andel G, van Helden SH, Boermeester MA (2010) Effect of a comprehensive surgical safety system on patient outcomes. N Engl J Med 363(20):1928–1937

de Vries EN, Ramrattan MA, Smorenburg SM, Gouma DJ, Boermeester MA (2008) The incidence and nature of in-hospital adverse events: a systematic review. Qual Saf Health Care 17(3):216–223

Francis R (2012) The Mid-Staffordshire NHS Foundation Trust public enquiry. http://www.mid-staffspublicinquiry.com/report

Gaba DM (2000) Structural and organizational issues in patient safety: a comparison of health care to other high-hazard industries. Calif Manage Rev 43(1)

Graber ML (2013) The incidence of diagnostic error in medicine. BMJ Qual Saf 22(Suppl 2): ii21–ii27

Haynes AB, Weiser TG, Berry WR, Lipsitz SR, Breizat AHS, Dellinger EP, Gawande AA (2009) A surgical safety checklist to reduce morbidity and mortality in a global population. N Engl J Med 360(5):491–499

Hollnagel E (2014) Safety-I and safety-II: the past and future of safety management. Ashgate Publishing, Ltd, Farnham, England.

Landrigan CP, Parry GJ, Bones CB, Hackbarth AD, Goldmann DA, Sharek PJ (2010) Temporal trends in rates of patient harm resulting from medical care. N Engl J Med 363(22):2124–2134

Lee LA, Domino KB (2002) The Closed Claims Project: has it influenced anaesthetic practice and outcome? Anesthesiol Clin North America 20(3):485–501

Michel P, Lathelize M, Quenon JL, Bru-Sonnet R, Domecq S, Kret M (2011) Comparaison des deux Enquêtes Nationales sur les Événements Indésirables graves associés aux Soins menées en 2004 et 2009. Rapport final à la DREES (Ministère de la Santé et des Sports)–Mars 2011, Bordeaux

Murphy MF, Waters JH, Wood EM, Yazer MH (2013) Transfusing blood safely and appropriately. BMJ 347:f4303

Pronovost P, Needham D, Berenholtz S, Sinopoli D, Chu H, Cosgrove S, Goeschel C (2006) An intervention to decrease catheter-related bloodstream infections in the ICU. N Engl J Med 355(26):2725–2732

Reason JT (1997) Managing the risks of organizational accidents, vol 6. Ashgate, Aldershot

Runciman W, Hibbert P, Thomson R, Van Der Schaaf T, Sherman H, Lewalle P (2009) Towards an International Classification for Patient Safety: key concepts and terms. International J Qual Health Care 21(1):18–26

Sevdalis N, Lyons M, Healey AN, Undre S, Darzi A, Vincent CA (2009) Observational teamwork assessment for surgery: construct validation with expert versus novice raters. Ann Surg 249(6):1047–1051

Sharpe VA, Faden AI (1998) Medical Harm: Historical, conceptual and ethical dimensions of iatrogenic illness. Cambridge University Press, Cambridge/New York

Singh H, Meyer A, Thomas E (2014) The frequency of diagnostic errors in outpatient care: estimations from three large observational studies involving US adult populations. BMJ Qual Saf. doi:10.1136/bmjqs-2013-002627

Vincent CA (1989) Research into medical accidents: a case of negligence? Br Med J 299(6708):1150–1153

Vincent C (2010) Patient safety. Wiley Blackwell, Oxford

Vincent C, Amalberti R (2015) Safety in healthcare is a moving target. BMJ Qual Saf. doi:10.1136/bmjqs-2015-004403

Vincent C, Moorthy K, Sarker SK, Chang A, Darzi AW (2004) Systems approaches to surgical quality and safety: from concept to measurement. Ann Surg 239(4):475–482

Vincent C, Burnett S, Carthey J (2013) The measurement and monitoring of safety. The Health Foundation, London

Vincent C, Burnett S, Carthey J (2014) Safety measurement and monitoring in healthcare: a framework to guide clinical teams and healthcare organisations in maintaining safety. BMJ Qual Saf 23(8):670–677

Wears RL, Hollnagel E, Braithwaite J (eds) (2015) Resilient health care, vol 2, The resilience of everyday clinical work. Ashgate Publishing, Ltd, Guildford

World Bank. Health expenditure total (% of GPD). http://data.worldbank.org/indicator/SH.XPD.TOTL.ZS?page=1. Accessed 1 Aug 2015

The Ideal and the Real

2

In this chapter we first attempt to persuade (or remind) the reader that much health-care departs from the care envisaged by standards and guidelines. We appreciate that standards and guidelines need considerable interpretation and adaptation for patients with multiple conditions (Tinetti et al. 2004) and that even the simplest conditions require consideration of personal preferences and other factors. However we are concerned primarily with the basic fact that the care provided to patients often does not reach the standard that professionals intend to deliver and which professional consensus would regard as reasonable and achievable. Clinical processes and systems are often unreliable and in fact many patients are harmed by the healthcare intended to help them. All this is to some degree obvious to anyone who works at the frontline or studies healthcare deeply. One of the questions we address in this book is how to manage the gap between the 'real and the ideal' and how best to manage the risks to patients.

Many factors conspire to make optimal care both difficult to define and difficult to achieve (Box 2.1). The vulnerabilities of the system, personal attitudes, team dynamics and a variety of external pressures and restraints combine to produce a 'migration' away from best practice. This in turn means that clinical staff are engaged in constant adaptation, detecting problems and responding to them. Safety is in a very real sense achieved by frontline practitioners rather than imposed by standards. We will develop this further in later chapters to argue that safety strategies to manage these risks need to foster these adaptive capacities both at an individual and organisational level.

© The Author(s) 2016
C. Vincent, R. Amalberti, *Safer Healthcare: Strategies for the Real World*,
DOI 10.1007/978-3-319-25559-0_2

Box 2.1 Observation of Patients at Risk of Suicide: When Working Conditions Make It Difficult to Follow Procedures

Over a 1 year period there were on average 18 suicides by in-patients under observation per year in hospitals in the United Kingdom. Ninety-one percent of deaths occurred when patients were under level 2 (intermittent) observation.

Deaths under observation tended to occur when policies or procedures (including times between observations) were not followed, for example:

- When staff are distracted by other events on the ward
- At busy periods, such as between 7.00 and 9.00
- When there are staff shortages
- When ward design impedes observation.

National Confidential Inquiry into Suicide and Homicide by People with Mental Illness (2015)

The Day-to-Day Realities of Healthcare

When we are working we are usually preoccupied with the task in hand and do not have the attentional capacity to simultaneously reflect on the working environment or remember all the difficulties encountered during the day. Furthermore, we are not easily able to aggregate our experience over long time periods. For instance, a doctor may know that notes are often missing in the clinic but will struggle to estimate how often this happens over a year. In addition it is very difficult for individuals to gain a true understanding of the failures and vulnerabilities across an entire technical area. Patients and families have a privileged view in that they alone follow the full story of care but our view as patients is obviously partial in that we cannot know the wider workings of the hospital or clinic. All these factors combine to make it difficult for any individual to monitor or assess the overall standard of care. There is however ample evidence to support the simple idea that care often falls below the standard expected. Let us consider some examples.

Comparing Actual Care with the Care Intended by Guidelines

Major studies in both the United States and Australia suggest that patients typically receive only a proportion of the care indicated by guidelines. Studies in the United States suggest that many patients received only about half of recommended care, though other patients receive investigations and treatment that are unnecessary (McGlynn et al. 2003). In a major recent Australian study adult patients received only 57 % of recommended care with compliance ranging from 13 % for alcohol dependence to 90 % for coronary artery disease (Runciman et al. 2012). These studies did not assess the direct impact on the patients concerned, but other studies have linked failures in the care provided with subsequent harm. For instance Taylor and colleagues (2008) interviewed 228 patients during and after their treatment and found 183 service quality deficiencies, each of which more than doubled the risk of any adverse event or close call for that patient. Service quality deficiencies involving poor coordination of care were particularly associated with the occurrence of adverse events and medical errors. In another example physicians reviewed 1566 case notes from 20 English Hospitals writing judgment-based comments on the phases of care provided and on care overall. About a fifth of the patients were considered to have received less than satisfactory care, often experiencing a series of adverse events (Hutchinson et al. 2013).

Reliability of Clinical Systems in the British NHS

Some healthcare processes, such as the administration of radiotherapy, achieve very high levels of reliability. Other processes are haphazard to say the least. Burnett and colleagues (2012) examined the reliability of four clinical systems in the NHS: clinical information in surgical outpatient clinics, prescribing for hospital inpatients, equipment in theatres, and insertion of peripheral intravenous lines. Reliability was defined as 100 % fault free operation when, for example, every patient had the required information available at the time of their appointment.

Reliability was found to be between 81 and 87 % for the systems studied, with significant variation between organisations for some systems; the clinical systems therefore failed on 13–19 % of occasions. This implies, if these findings are typical, that in an English hospital: doctors are coping with missing clinical information in three of every 20 outpatient appointments and there is missing or faulty equipment in one of seven operations performed. In each case where measured, about 20 % of reliability failures were associated with a potential risk of harm. On this basis it is hardly surprising that patient safety is routinely compromised in NHS hospitals and that clinical staff come to accept poor reliability as part of everyday life.

Following the Rules: Reliability of Human Behaviour

Delivering safe, high quality care is an interplay between disciplined, regulated behaviour and necessary adaptation and flexibility. Rules and procedures are never a complete solution to safety and sometimes it is necessary to depart from standard procedures in the pursuit of safety. However, protocols for routine tasks are standardised and specified precisely because those tasks are essential to safe, high quality care.

Protocols of this kind are equivalent to the safety rules of other industries – defined ways of behaving when carrying out safety-critical tasks (Hale and Swuste 1998). Examples in healthcare include: checking equipment, washing your hands, not prescribing dangerous drugs when you are not authorised to, following the procedures when giving intravenous drugs and routinely checking the identity of a patient. Such standard routines and procedures are the bedrock of a safe organisation, but there is ample evidence that such rules are routinely ignored:

- Hand washing. Contamination through hand contact is a major source and hand hygiene a major weapon in the fight against infection (Burke 2003). Studies have found that average levels of compliance, before major campaigns were insti-tuted, have varied from 16 to 81 % (Pittet et al. 2004).

- Intravenous drug administration. Studies have found that over half involve an error, either in the preparation of the drug or its administration. Typical errors were pre-paring the wrong dose or selecting the wrong solvent (Taxis and Barber 2003).
- Prophylaxis against infection and embolism. Only 55 % of surgical patients receive antimicrobial prophylaxis (Bratzler et al. 2005) and only 58 % of those at risk of venous thromboembolism receive the recommended preventive treatment (Cohen et al. 2008).

The causes of departure from standards are many. In some settings the working environment is reasonably calm and orderly so staff are able to follow clear proto-cols and abide by core standards. In other settings however the pressures are great, the environment noisy and chaotic and staff are essentially just trying to do the best they can in the circumstances. In any systems there are pressures for greater produc-tivity, less use of resources and occasions where missing or broken equipment forces adaptations and short cuts; add to this that we all, occasionally or frequently, are in a rush to get home, get on to the next case, tired or stressed and apt to cut corners. Standards may be unrealistic or too complex; staff may not be sufficiently skilled or have not received the necessary training. Working in such conditions is an everyday occurrence for many clinicians and acts as a constant reminder of the care they would like to give and the reality of the care they are able to provide. Over time however these departures from standards can become increasingly tolerated and eventually invisible (Box 2.2).

Box 2.2 External Pressures and Gradual Migration to the Boundary of Safety
Occasional lapses can become more tolerated over time and systems can become gradually more degraded and eventually dangerous. The phrase 'ille-gal normal' captures the day-to-day reality of many systems in which devia-tions from standard procedures (the illegal) are widespread but occasion no particular alarm (they become normal). The concept of routine violations is not part of the thinking of managers and regulators; in truth it is a very uncom-fortable realisation that much of the time systems, whether healthcare, trans-port or industry, operate in an 'illegal-normal' zone. The system continues in this state because the violations have considerable benefits, both for the indi-viduals concerned and for managers who may tolerate them, or even encour-age them, in the drive to meet productivity standards.

Over time these violations can become more frequent and more severe so that the whole system 'migrates' to the boundaries of safety. Violations are now routine and so common as to be almost invisible to both workers and managers. The organisation has now become accustomed to operating at the margins of safety. At this stage, any further deviance may easily result in patient harm, and would generally be considered as negligent or reckless con-duct (Amalberti et al. 2006).

The Ideal and the Real: Five Levels of Care

We now consider the implications of the gap between the care envisaged by standards and guidelines and the care actually given to patients. We have found it useful to distinguish five levels of care each departing further from the ideal and, we suggest, increasing probability of harm as one moves down the levels.

1. Level 1 corresponds to optimal care envisaged by standards (though truly optimal care can never be encapsulated in standards). These standards are set out by national and professional organisations and represent a consensus on what can be regarded as the optimum care achievable within current cost constraints. This level provides a shared ideal reference of excellent care, although it is seldom fully achieved across an entire patient journey.
2. Level 2 represents a standard of care which experts would judge as both providing a good outcome for the patient and also achievable in day to day practice. The care is of good standard and the outcome is good, even though there may be minor variations and problems. Any departures from best practice are relatively unimportant in the overall care provide to the patient.
3. Level 3 represents the first level where the safety of the patient may be compromised. We consider, for reasons given above, that a considerable amount of the healthcare that patients receive falls broadly into this category. At this level there are frequent departures from best practice which occur for a wide variety of different reasons and are a potential threat to patients. There may for example not be a timely monitoring of anticoagulation level after prescription of heparin. This level has been previously described as the 'illegal normal' (Amalberti et al. 2006) (Box 2.2).
4. Level 4 represents a departure from standards which is sufficient to produce avoidable harm. For example, a 68 year old patient undergoes a cholecystectomy and contracts a urinary catheter infection after surgery. Analysis of the event showed that the catheter was not checked regularly and was left in place too long. This was a clear departure from expected care. However treatment was rapidly instituted and the infection was under full control after 10 days. The patient suffered avoidable harm and had to stay in hospital an additional week but then recovered completely.
5. Level 5 refers to care that is poor over a longer period and places the patient at risk of substantial and enduring harm. For instance if, in the case described above, the patient not only contracted the infection but it was then not recognised and not treated effectively. This would result in at best a very prolonged recovery and increased frailty but also a potentially fatal outcome.

Broadly speaking we see Level 1 as optimal care, certainly a valuable aspiration and inspiration but very difficult to achieve in practice and in many settings not easy to define. Optimal care is relatively easy to specify in a highly standardised and structured clinical setting but in many environments, particularly primary care, the care provided necessarily evolves and unfolds in a complex social context (Box 2.3). Level 2 is a more realistic level of care where there are minor imperfections but clinical care is of a very good standard. Level 3 is a distinct deterioration with multiple lapses of care but not sufficient to greatly affect long term outcome. Levels 4 and 5 in contrast offer potential for harm, either through omission of critical aspects of care, serious errors or neglect. In the common understanding of these terms the ambitions of high quality care are associated with Levels 1 and 2, and those concerned with safety aiming to avoid levels 4 and 5.

Box 2.3 Optimal Care Can Often Not Be Precisely Defined

There are many clinical situations in which optimal care cannot be precisely defined. This may be because the disease is not well understood, is rare or expressed in an unusual manner. More commonly though the patient, often frail or elderly, is suffering from a number of different conditions presenting a complicated and changing picture. In these cases, common in primary care and mental health, clinical judgement and sensitive shared decisions are to the fore:

> I don't believe that in much of what we do in healthcare, particularly in primary care, we can define what we mean by 'excellence', nor can we codify it though guidelines and standards. We can reduce health provision to its component parts and pretend that these reflect the whole but this ignores the inherent paradoxes of competing goods and trade-offs with other objectives. The thinking that dominates the safety world is sometimes too rational. What I see is lots of thoughtful clinicians who understand the discrepancy between the ideal and the real, for whom the tension is always on their agenda and who thoughtfully manage these tensions because they accept that they live in a world that wants to simplify (M Marshall, 2015, personal communication)

Figure 2.1 is similar to many diagrams which represent variation in standards of care and which distinguish good, average and poor units or organisations. Certainly some organisations deliver poor care for sustained periods of time and even national services can have periods of high risk of harm at times of crisis. However we intend to capture a more fluid reality in which any patient is at risk of a sudden decline in standards, and at risk of harm, on many occasions during their healthcare journey. Safety can be eroded quite suddenly in any team or organisation, just as there is always some risk of accident with the safest car driven by the best driver on the safest road. We are therefore not only concerned with strategies which may support struggling teams or organisations but also with developing strategies and interventions to manage risk on a day-to-day basis.

The Cumulative Impact of Poor Quality Care

Patients can receive some treatment of poor quality, in the sense of haphazard and patchy adherence to accepted standards, and still not come to any harm. We suggest, however, harm is much more probable when healthcare moves further from best

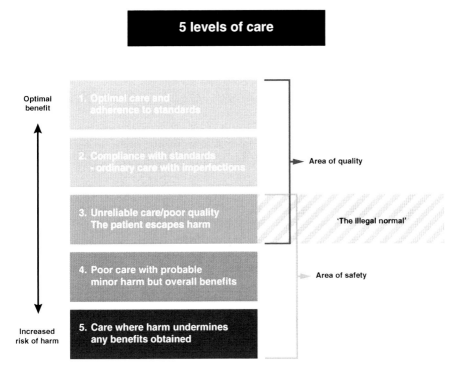

Fig. 2.1 Five levels of care

practice (Levels 4 and 5). This is partly because obvious lapses in standards (such as not checking patient identity) may lead to immediate harm but the greater danger to patients probably comes from the cumulative impact of minor problems (Hutchinson et al. 2013).

Suppose for example, a fit and well 26 year old patient has planned abdominal surgery for inflammatory bowel disease on a Wednesday. Due to a shortage of beds, the patient is placed on the orthopaedic ward, with nurses, pharmacists and other clinicians who are not used to looking after this type of problem. The operation is difficult and complex; a piece of bowel was removed and a new join made between the remaining ends of the bowel. On Saturday evening the patient has an episode of fever at 39 °C and abdominal pain, which could indicate that the new join is leaking. The young doctor on duty over the weekend is a locum, who does not know the patient. She tries to read the operation note but it is not completely legible. She does not appreciate the potential seriousness and does not seek more senior advice. Over the next 24 h the patient continues to deteriorate but the staff do not appreciate the significance of the symptoms. By Monday morning the patient is so unwell that he suffers a cardiac arrest and eventually dies on Monday evening after a futile return to the operating theatre during the day.

This scenario describes a series of relatively ordinary and commonplace lapses, omissions and events which together have a catastrophic effect. Clearly there were problems in assessment of symptoms, escalation, record keeping, communication, coordination of care and management of bed availability, possibly exacerbated by external pressures. None of the individual lapses and problems is out of the ordinary or particularly shocking – but they combine to create catastrophe.

The impact of the cumulative effects of poor care suggests that we may need to widen the time frame of analyses of adverse events and poor outcomes. This will be especially important once we consider safety in the home and community and in the context of the overall impact of healthcare on a person's life and well-being. However we may also see a much greater incidence of harm due to cumulative minor failure in the future owing to the number and complexity of transitions along the patient journey.

Consideration of the cumulative effects of poor care also has implications for how we assess priorities for patient safety. Dramatic incidents, such as deaths from spinal injection of vincristine (Franklin et al. 2014), attract considerable attention and are tragic for the people involved but they tend to skew the direction of patient safety initiatives towards comparatively rare events. In surgery the cases that attract most attention are those with sudden, dramatic outcomes with fairly immediate causes. These are incidents such as operating on the wrong patient or retained foreign body which are rare but frequently disastrous when they do occur; they are low risk but 'high dread' in the language of the psychology of risk. However surgical patients run much greater risks from care that is simply of poor standard for whatever reason. There is for instance a huge variation in mortality from surgery across Europe. In a major recent review 46,539 patients were studied of whom 1855 (4 %) died before hospital discharge (Pearse et al. 2012). Crude mortality rates varied widely between countries range from 1.2 % for Iceland to 21.5 % for Latvia. Substantial differences remained even after adjustment for confounding variables. This suggests that much of the care provided is, in our terms, of levels 3, 4 and 5 even by the standards of individual countries. Once we begin to see safety in these terms it is clear that the harm from poor management of post-operative complications dwarfs the much more prominent problem of surgical never events.

We need therefore to reflect on the broader priorities from a population health perspective. This process has already begun with the increased attention given to programmes to reduce falls, pressure ulcers, acute kidney injury and infections of all kinds. The scope of patient safety needs to further expand to embrace consideration of poor care of all kinds and to integrate with those seeking to understand and reduce the sources of variability. We should also remember that most studies at the hospital level focus on one particular type of adverse event (such as hospital acquired infection) or one service (such as surgery). Very few studies assess the whole spectrum of incidents afflicting patients or assess their cumulative impact over time.

Explicit Discussion of the Real Standard of Care Is Critical

We now come to a central problem and challenge which is that the standard of usual care cannot easily be explicitly discussed. It is of course understood implicitly within clinical teams and each new member is socialised into accepting the standards of care in that particular environment which may be either higher, or lower, than they are used to. When people join a unit, or spend time in another unit, there may be a sudden shock of recognition of a very different standard and tolerance for departures from standards. There is huge variation in different clinical teams in what they regard as good enough care which is influenced by the social norms and values of that particular setting. Care that is viewed as ordinary on one ward might be seen as being a major lapse in standards on another. Most clinicians are aware that much care is in the 'illegal normal' range and immediately recognise this concept when it is presented. They know that the care they provide often falls short of the care they would like to provide but they are adept at navigating the healthcare system to provide the best care they can in the circumstances.

Organisations however, and still more governments, cannot easily openly say that much care is at level 3 and routinely dips to levels 4 and 5. This has some important consequences for the management of risk. First, it becomes very difficult to study or to value the many adaptive ways in which staff cope in difficult environment to prevent harm coming to patients. Second, and most important to our arguments, attempts to improve safety may not be targeting the right levels or the right behaviours. We will argue later that most safety interventions are essentially attempts to improve reliability and, ultimately, to move all care towards Level 1. This is an important and necessary strategy but, in our view, only applicable in some circumstances. This approach need to be supplemented by strategies that aim to maintain care at Level 3 and prevent decay into levels 4 and 5. We might express this by saying that in the day to day provision of care it is more urgent that our systems prioritise achieving reliable basic standards than striving for unachievable ideals. If care is generally at Level 3 then the principal aim might be to improve reliability and move to Level 2. If however care is often at level 4 or 5, that is frankly dangerous, then the detection and response to potential dangers might be a higher priority (Fig. 2.2).

The aspirations to excellence are important and should not be mocked or derided as unrealistic. The problem is that the rhetoric of excellence masks the urgently needed discussion of the realities of 'usual care'. If our aspiration becomes only to deliver 'good enough' care then there is a danger of increased variation, declining standards and increased hazard. The definition and aspiration of optimum quality remains critical – but so does an explicit discussion of the current reality.

What Is the Impact of Improving Quality Standards?

Innovation and the aspiration to continually improve are at the heart of medicine. However the introduction of new treatments or new standards of care may place unrealistic demands on both staff and organisations. Stroke for example was at one time regarded as untreatable. Brain cells were thought to die within minutes after a stroke began, and medical treatment largely consisted of caring for the patient and "wait and see". We have known for a decade now that treatment following a stroke, especially if begun within 3 h of onset, can preserve brain tissue. Guidelines typically recommend a door-to-needle time of 60 min. However in 2011 only one third of American patients were treated within the guideline-recommended door-to-needle times (Fonarow et al. 2013). Many countries have instituted major programmes to improve the efficiency of treatment for stroke which have led to great improvements in outcome.

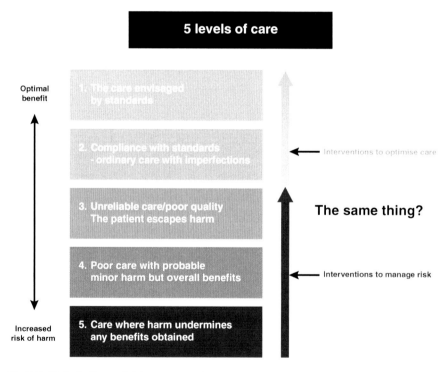

Fig. 2.2 Optimisation and risk management

We are not criticising the care given or the delays in bringing in new standards of delivery. Rather we are pointing to the inevitable increase in departures from guidelines which result when a new standard is specified. Ten years ago there would have been few 'incidents' relating to failures in the early treatment of stroke because standards of care had not been introduced. We can now, because of improved care, point to numerous serious incidents because many patients cannot access to care within the 3 h of onset due to a failures in the healthcare system. As standards improve we are therefore likely to have an increasing number of incidents which are concerned with omissions of care. What counts as an 'incident' in 2015 may simply have been ordinary practice in 2005; this is a very common consideration in legal cases which are being decided some years after the initial event. The new standards, hugely beneficial for patients, create new kinds of incidents and safety problems (Vincent and Amalberti 2015).

Levels of Care and Strategies for Safety Improvement

Improving the standard of care delivered and the gradual setting of higher standards is of course a positive and necessary aspiration. However when doing this we need to recognise that we are redefining both quality and safety and increasing the pressure on individuals and organisations. Champions of the new standards will emerge and bring about change but many organisations will take time to meet the new standard and weaker organisations may even be destabilised because of the increasing demand.

We have at present very few strategies for managing the transitional period or for responding constructively to the inevitable gap between expected standards and organisational reality. During these transition periods we will need to do more than simply exhort and harass organisations to meet the new standards. We must also recognise the inevitable lag and employ strategies that emphasise the detection of problems, awareness of conditions which degrade safety and individual and enhance team based management of potentially harmful care. These arguments will be fully developed in Chap. 7.

Key Points

- Many patients do not receive the care intended. We can do a great deal to increase reliability and achieve higher standards of care. However we believe that in healthcare there will always be a gap between the ideal and the real.
- We distinguish five levels of care each departing further from the ideal and, we suggest, increasing probability of harm as one move down the levels.
  - Level 1 corresponds to the optimum care envisaged by standards. This level provides a shared reference of excellent care, although it is seldom fully achieved across an entire patient journey.
  - Level 2 represents a standard of care which experts would judge as both providing a good outcome for the patient and also achievable in day to day practice
  - Level 3 represents the first level in which the safety of the patient is threatened. At this level there are frequent departures from best practice which occur for a wide variety of different reasons.
  - Level 4 represents a departure from standards which is sufficient to produce avoidable harm but not sufficient to substantially affect the overall outcome.
  - Level 5 refers to care that is poor over a longer period and places the patient at risk of substantial and enduring harm.
- We suggest that organisations and government find it difficult to openly discuss the daily threats and variations in standards of care. This has important consequences. First, it becomes very difficult to study or value the many ways staff adapt to prevent harm coming to patients. Second, attempts to improve safety may not be targeting the right levels or the right behaviours.
- The aspirations to excellence are important and should not be derided as unrealistic. The problem is that the rhetoric of excellence masks the urgently needed explicit discussion of the realities of usual care which is a critical first step in the effective management of risk.
- We propose that most safety interventions are essentially attempts to improve reliability. This is an important and necessary approach but needs to be complemented by additional strategies that aim to manage risk and protect patients from serious failures in care.

References

Amalberti R, Vincent C, Auroy Y, de Saint Maurice G (2006) Violations and migrations in health care: a framework for understanding and management. Qual Saf Health Care 15(suppl 1):i66–i71

Bratzler DW, Houck PM, Richards C, Steele L, Dellinger EP, Fry DE, Red L (2005) Use of anti-microbial prophylaxis for major surgery: baseline results from the National Surgical Infection Prevention Project. Arch Surg 140(2):174–182

Burke JP (2003) Infection control-a problem for patient safety. New England Journal of Medicine, 348(7):651–656

Burnett S, Franklin BD, Moorthy K, Cooke MW, Vincent C (2012) How reliable are clinical systems in the UK NHS? A study of seven NHS organisations. BMJ Qual Saf. doi:10.1136/bmjqs-2011-000442

Cohen AT, Tapson VF, Bergmann JF, Goldhaber SZ, Kakkar AK, Deslandes B, Huang W, Zayaruzny M, Emery L, Anderson FA Jr, Endorse Investigators (2008) Venous thromboembolism risk and prophylaxis in the acute hospital care setting (ENDORSE study): a multinational cross-sectional study. Lancet 371(9610):387–394

Fonarow GC, Liang L, Smith EE, Reeves MJ, Saver JL, Xian Y et al (2013) Comparison of performance achievement award recognition with primary stroke centre certification for acute ischemic stroke care. J Am Heart Assoc 2(5), e00045

Franklin BD, Panesar S, Vincent C, Donaldson L (2014) Identifying systems failures in the pathway to a catastrophic event: an analysis of national incident report data relating to vinca alkaloids. BMJ Qual Saf 23(9):765–772

Hale AR, Swuste PHJJ (1998) Safety rules: procedural freedom or action constraint? Saf Sci 29(3):163–177

Hutchinson A, Coster JE, Cooper KL, Pearson M, McIntosh A, Bath PA (2013) A structured judgement method to enhance mortality case note review: development and evaluation. BMJ Qual Saf 22(12):1032–1040

McGlynn EA, Asch SM, Adams J, Keesey J, Hicks J, DeCristofaro A, Kerr EA (2003) The quality of health care delivered to adults in the United States. N Engl J Med 348(26):2635–2645

National Confidential Inquiry into Suicide and Homicide by People with Mental Illness (NCISH) (2015) In-patient suicide under observation. University of Manchester, Manchester

Pearse RM, Moreno RP, Bauer P, Pelosi P, Metnitz P, Spies C, Rhodes A (2012) Mortality after surgery in Europe: a 7 day cohort study. Lancet 380(9847):1059–1065

Pittet D, Simon A, Hugonnet S, Pessoa-Silva CL, Sauvan V, Perneger TV (2004) Hand hygiene among physicians: performance, beliefs, and perceptions. Ann Intern Med 141(1):1–8

Runciman WB, Hunt TD, Hannaford NA, Hibbert PD, Westbrook JI, Coiera EW, Braithwaite J (2012) CareTrack: assessing the appropriateness of health care delivery in Australia. Med J Aust 197(10):549

Taxis K, Barber N (2003) Causes of intravenous medication errors: an ethnographic study. Qual Saf Health Care 12(5):343–347

Taylor BB, Marcantonio ER, Pagovich O, Carbo A, Bergmann M, Davis RB, Weingart SN (2008) Do medical inpatients who report poor service quality experience more adverse events and medical errors? Med Care 46(2):224–228

Tinetti ME, Bogardus ST Jr, Agostini JV (2004) Potential pitfalls of disease-specific guidelines for patients with multiple conditions. N Engl J Med 351(27):2870–2874

Vincent C, Amalberti R (2015) Safety in healthcare is a moving target. BMJ Qual Saf. doi:10.1136/bmjqs-2015-004403

Approaches to Safety: One Size Does Not Fit All

3

In the previous chapter we set out five levels of care with the levels being defined according to how closely they met expected standards of care. We argued that the care delivered to patients frequently departs from expected standards and that this has important implications for the management of safety. Most safety improvement strategies aim to improve the reliability of care and move more closely to optimal care. We suggest that these strategies need be complemented by strategies that are more concerned with detecting and responding to risk and which assume that care will often be delivered in difficult working conditions.

This argument could be seen simply as an admission of defeat. We might appear to be saying that healthcare will never achieve the safety standards of commercial aviation and we must accept this and manage the imperfections as best we can. Errors will inevitably occur, patients will sometimes be harmed and the best we can hope for is to respond quickly and minimize the damage. We would accept that working conditions and levels of reliability are often unnecessarily poor and that strategies to manage these risks to patients are much needed. However, there are more fundamental reasons for widening our view of safety strategies beyond trying to improve reliability. The more critical point is that different challenges and different types of work require different safety strategies. One safety size does not fit all.

Approaches to Risk and Hazard: Embrace, Manage or Avoid

The metaphor of the climber and the rock face serve as a framework to introduce our discussion of contrasting approaches to safety. Hazards in healthcare are like rock faces for climbers, an inevitable part of daily life. These hazards have to be faced but this can be done in very different ways. One can minimize the risk by refusing to climb unless conditions are perfect (plan A). Alternatively one can accept higher

© The Author(s) 2016
C. Vincent, R. Amalberti, *Safer Healthcare: Strategies for the Real World*,
DOI 10.1007/978-3-319-25559-0_3

levels of risk but prepare oneself to manage the risk effectively. A climber or team of climbers may attempt a dangerous rock face but only after careful preparation, establishing clear safety procedures and plans for dealing with emergencies. A well prepared and coordinated team can achieve much higher levels of safety than an individual (plan B). Finally, a climber may simply embrace risk and rely on personal skill and resilience to deal with whatever occurs. They may climb without proper equipment, without training or in deteriorating weather conditions; or more dangerous still they may dare to climb unknown rock faces taking massive risks in the spirit of personal challenge and competitive achievement (plan C). All these climbers are concerned with safety but they vary in the risks they are prepared to take and the strategies they adopt (Amalberti 2013).

Some professions, such as fighter pilots, deep sea fishing skippers and professional mountaineers, literally make a living from exposure to risk. In these professions, accepting risk, and even seeking out risk, forms the essence of their work. These professions do, however, still want to improve safety. A number of studies carried out among fighter pilots (Amalberti and Deblon 1992) and sea fishing skippers (Morel et al. 2008, 2009) show that they have a real desire for safety. Fishing skippers, for example, would like to have an intelligent anti-collision system to offer them better protection in high seas and with poor visibility which would give increased mobility for trawling. Fighter pilots would like an electronic safety net to offer them better protection when they are undertaking manoeuvres that may cause them to lose consciousness.

People who rely on their personal skill and resilience are not reckless; a few may be but they are not likely to survive long. They usually have a core set of safety procedures that they take very seriously. The problem is that the constantly changing environment in which they work does not lend itself to managing risk by using rules and procedures. (If they did, one would change to a plan B approach). Instead, the response is necessarily ad-hoc because the environment is constantly changing and because economic considerations often drive people to take greater risks. Plan C solutions are essentially resilient in character: becoming more expert, becoming able to judge the difficulty of the task, being realistic about one's own skills and acquiring experience which allows adaptation to uncertain or dangerous conditions.

In contrast, the high levels of safety in civil aviation are achieved by very different means. Here, the solution is radically different and frequently involves not exposing crews to the hazardous conditions that increase the risk of accidents. For example, the eruption of the Eyjafjallajökull volcano in Iceland in 2010 led to all European aircraft immediately being grounded based on a simple approach: no unnecessary exposure to risk. Deep sea fishing and commercial aviation reveal contrasting strategies for dealing with risk. The first, typical of very competitive and dangerous activities, involves relying on the intelligence and resilience of frontline operators and giving them aids to deal with risk; the other relies on organisation, control and supervision and ensures that operators are not exposed to risks. Both of these models take safety very seriously but they manage risk in very different ways (Fig. 3.1).

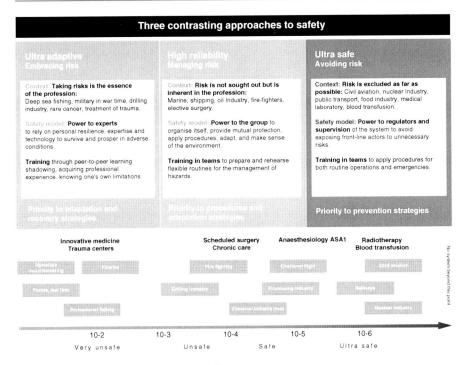

Fig. 3.1 Three contrasting approaches to safety

Three Approaches to the Management of Risk

We can then distinguish three broad approaches to the management of risk each with its own characteristic approach. Each one has given rise to an authentic way of organising safety with its own characteristic approach and its own possibilities for improvement (Grote 2012; Amalberti 2013). In practice the distinctions may not be that clear cut but the division into three models serves to illustrate the principal dimensions and factors in play (Box 3.1).

Box 3.1 A Note on Terminology

We have chosen three terms to describe contrasting approaches to safety: 'ultra-adaptive', high reliability and 'ultra-safe'. All of these terms, particularly the first two, are associated with a number of theories and concepts. In this book we use these terms in a more descriptive sense. 'Ultra-adaptive' simply means that this approach relies heavily on the judgement, adaptability and resilience of individuals; 'high reliability' does indeed reflect the literature on high reliability organisations (HROs), but here is mainly meant to indicate a flexible but prepared response of teams in the management of risk; 'ultra-safe' refers to the absolute priority safety has in those environments and to the means of achieving such safety.

Embracing Risk: The Ultra-adaptive Model

This approach is associated with professions in which seeking exposure to risk is inherent in the activity and often also embedded in the economic model of that profession. Skilled professionals sell their services on the basis of their expertise and willingness to embrace risk, master new contexts, cope and win through, reaping benefits where others fail or are afraid to go. This is the culture of champions and winners, and there are of course those who fail to meet the challenges or who are injured or die in the attempt. This tends to be explained in personal terms; they did not have the knowledge or skill of the champions; they did not have the 'right stuff' to be part of these elite groups (Wolfe 1979). Deep sea fishing skippers, for example, are willing to seek out the riskiest conditions in order to catch the most profitable fish at the best times. Such professions are very dangerous and have appalling accident statistics. They are not, however, insensitive to the risks they run. They have safety and training strategies which are very well thought-out, but they are highly reliant on individual skills and strongly influenced by their own particular culture.

Within the ultra-adaptive model individual autonomy and expertise take precedence over the hierarchical organisation of the group. In many cases the group is very small (consisting of two to eight individuals) and works in a highly competitive environment. The leader is recognised for technical ability, past performance and charisma more than his official status. Everyone involved has to use a high degree of initiative. Skill, courage and accumulated experience combined with a clear-eyed awareness of personal strengths and limitations are the keys to recognition as a good professional. Success is seen in terms of winning and surviving, and only winners have a chance to communicate their safety expertise in the form of champions' stories.

To summarise, there are a small number of procedures, a very high level of autonomy and a very large number of accidents. Becoming more effective and learning to manage risk are achieved by working alongside experts, learning from experience and increasing one's own capacity to adapt and respond to even the most difficult situations. The differences between the least safe and the safest operators within a single resilient, skilled trade are of the order of a factor of 10; for instance the rate of fatal accidents in professional deep-sea fishing varies by a factor of 4 between ship owners in France and by a factor of 9 at the global level (Morel et al. 2009). This suggests that that it is certainly possible to make progress through safety interventions within this particular model of safety; there may however be a limit to what can be achieved without moving to a different model of safety which in turn would require a radically different approach to the activities concerned (Amalberti 2013).

Managing Risk: The High Reliability Approach

The term high reliability or high reliability organisation (HRO) is most associated with a series of studies of industries in which highly hazardous activities, such as nuclear power and aircraft carriers, were managed safely and reliably. A very wide variety of characteristics were identified as characterising high reliability organisations but all were underpinned by a disciplined but flexible approach to teamwork (Vincent et al. 2010). This approach also relies heavily on personal skill and resilience but in a more prepared and organized way; individual initiative must not come at the expense of the safety and success of the wider team (Weick and Sutcliffe 2007).

This approach is also associated with hazardous environments but the risks, while not entirely predictable, are known and understood. In these professions risk management is a daily affair, though the primary aim is to manage risk and avoid unnecessary exposure to it. Firefighters, the merchant navy, operating theatre teams, and those operating chemical factories all face hazards and uncertainty on a daily basis and typically rely on a high reliability model.

The HRO approach relies on leadership and an experienced professional team, which usually incorporates several different roles and types of expertise. All members of the group play a part in detecting and monitoring hazards (sense making), bringing them to the attention of the group, adapting procedures if necessary, but only when this makes sense within the group and is communicated to everyone. The HRO model is in fact relatively averse to individual exploits that are outside the usual repertoire of the team. The resilience and flexibility of approach employed is that of a dynamic and well-coordinated team rather than that of an individual acting on their own. All members of the group show solidarity in terms of safety objectives and the team promotes prudent collective decision-making.

The teams who work within this model place great importance on analysing failures and seeking to understand the reasons behind them. The lessons drawn from these analyses primarily concern ways in which similar scenarios could be managed better in future. This is therefore a model which relies firstly on improving detection and recovery from hazardous situations, and secondly on improving prevention – which means avoiding exposure to difficult situations when possible. Training is based on collective acquisition of experience. Once again, the differences between the best operators and those that are less good within a single trade are of the order of a factor of 10.

Avoiding Risk: The Ultra-safe Approach

With this approach we turn radically away from reliance on human skills and ingenuity towards a reliance on standardization, automation and avoidance of risk wherever possible. Professionalism in these contexts still requires very high levels of skill but the skills consists primarily in the execution of known and practiced routines, covering both routine operations and emergencies. Ideally, there is no need to rely on exceptional expertise even when dealing with emergencies, such as an engine fire on a commercial aircraft. Instead comprehensive preparation and training allows all operators to meet the required standards of performance and be skilled to the point that they are interchangeable within their respective roles.

This approach relies heavily on external oversight and the control of hazards which makes it possible to avoid situations in which frontline staff are exposed to exceptional risks. By limiting the exposure to a finite list of breakdowns and difficult situations, the industry can become almost completely procedural, both when working under normal conditions and under more difficult conditions. Airlines, the nuclear industry, medical laboratories and radiotherapy are all excellent examples of this category. Accidents are analysed to find and eliminate the causes so that exposure to these risky conditions can be reduced or eliminated in the future. Training of front-line operators is focused on respect for their various roles, the way they work together to implement procedures and how to respond in a prepared manner to any emergency, so that there is minimal need for improvised ad hoc procedures. Once again, the best and the least good operators within a single profession differ by about a factor of 10.[1]

Rules and Adaptation

We commonly assume that safety is achieved by imposing rules and restricting the autonomy of management and workers. We know however that writing a safety plan, specifying rules and compliance to legal requirements, offers no guarantee that the plan will be put into practice. There is a great deal of evidence about the extent of non-compliance to rules and safety standards and a recurring list of reasons for not adhering to the rules – too many, not understood, not known, not applying to non-standard cases, contradictions between rules and so on (Lawton 1998; Carthey et al. 2011). Moreover, workers in many organisations find that it is often necessary to by-pass or flout rules to get the job done in a reasonable time; these are 'optimising violations' in James Reason's memorable phrase (Reason 1997).

The three approaches to safety take radically different approaches to rules and procedures on the one hand and flexibility and adaptation on the other. Each approach has its own approach to training, to learning and improvement and each has its own advantages and its own limitations (Amalberti 2001; Amalberti et al. 2005). They can be plotted along a curve in which there is a trade-of between flexibility and adaptability on the one hand and standardisation and procedures on the other. It is important to acknowledge that while these approaches vary considerably in the way they manage risk all share the same ambition of being as safe as possible in the circumstances in which they operate.

Concrete safety results are therefore the product of apparent contradictory actions: rules and constraints that guide work on the one hand and reliance on the adaptive capacities of staff in scenarios that fall outside guidelines, rules and regulations. Staff sometimes fail to follow rules in a reckless or careless manner (i.e. for no good reason) but equally often the rules are ignored because they impede the

---

[1] The rate of aviation accidents ranges from 0.63 per million departures in Western countries to 7.41 per million departures in African countries. These therefore differ by a factor of 12, source: IATA statistics, 23 February 2011, http://www.iata.org/pressroom/pr/pages/2011-02-23-01.aspx

actual work itself. In healthcare we have the highly undesirable situation of a vast number of procedures and guidelines (far too many for staff to follow or even know about) which are followed inconsistently (Carthey et al. 2011). One critical task for healthcare in all settings, whether adaptive or standardised, is to identify a core set of procedures which really do have to be reliably followed.

How Many Models for Healthcare?

For the sake of simplicity we have viewed each industry as primarily being associated with one particular model. The reality is more complicated. For instance the activity of drilling for oil necessarily involves embracing risks; oil processing on the other hand, while hazardous, can potentially be managed in a way that minimises risk.

Healthcare is a particular complex environment. We have already alluded to this when commenting that healthcare is 'twenty different industries'. Consider the hospital environment with multiple types of work, many different professions and varying working conditions across clinical environments. There are areas of highly standardized care which conform very closely to our ultra-safe model. These include pharmacy, radiotherapy, nuclear medicine and much of the process of blood transfusion. All of these are highly standardized and rely heavily on automation and information technology. They are islands of reliability within the much more chaotic wider hospital environment. In contrast much ward care corresponds to our intermediate model of team based care where standards and protocols provide important controls on hazards (such as infection from poor hand hygiene) but where professional judgement and flexibility is essential to providing safe, high quality care.

Other sections of the hospital, such as emergency surgery, deal continually with complex cases and have to work in very difficult conditions. The work may be scheduled but there is considerable hour-to-hour adaptation due to the huge variety of patients, case complexity, and unforeseen perturbations. We should emphasise though that all clinical areas, no matter how adaptive, rely on a bedrock of core procedures; adaptive is a relative term not an invitation to abandon all guidelines and go one's own way. In addition, much clinical activity could be much more controlled than is often the case. In many hospitals elective and emergency surgery are still carried out by the same teams on the same day which ensures constant disruption to the routine procedures and insufficient focus on emergency patients, moving the whole system to a highly adaptive mode. Separating elective and emergency work and allocating separate teams to deal with each allows both areas to operate in a largely high reliability mode.

All of these professional activities have to adapt to changing staffing patterns and other pressures on the system. On 'Tuesday morning' (optimum working conditions) it may be possible for an emergency surgery team to adopt the characteristics of a HRO system. In contrast on 'Sunday night' (short staffed, lack of senior staff, lack of laboratory facilities) the team is forced to rely on a more adaptive approach. Healthcare is a wonderful arena for the study of safety, probably much better than any other setting, because the entire range of approaches and strategies can be found within one industry.

Moving Between Models

We sometimes assume that the safety ideal is the ultra-safe model of commercial aviation and other highly standardized processes. In a sense this is correct, in that this model is indeed very safe, but we have argued that this model is only workable with very specific conditions and strong constraints on risky activity. The model may not be appropriate, or even feasible, in other settings. Nevertheless in certain activities we can identify a move between different models according to circumstances.

The case of fighter pilots is a special and interesting case of a dual context: in peacetime, the air force requires them to operate in an essentially ultra-safe mode, but once the aircraft are deployed on active service, the operating model immediately becomes one of adaptation and resilience. The switch between these modes of operation can generate surprises in both directions. After returning from military campaigns pilots can persist in resilient and deviant behaviour as they struggle to return to peace time conditions. Conversely, when pilots are suddenly thrust from peacetime into operational theatre, important opportunities can be missed during the first few days of military engagement due to lack of practice in the resilient model.

Surgery offers similar parallels in that different forms of surgery correspond to different models and the same surgeons may need to adapt to different approaches. Highly innovative surgery, such as early transplant surgery, or surgery conducted in unusual environments such as field hospitals, clearly requires a risk embracing highly adaptive approach. The phrase 'heroic surgery' speaks exactly to this kind of intervention with the allusions to the personal qualities of the surgeon that are required, although greater heroism is probably required of the patient. We can see also that patients may also choose strategies which are very risky but yet are justified by the severity of the illness and the potential benefit.

Over time certain kinds of surgery may evolve through different models, beginning as a very high risk experimental procedure, moving towards a stage in which risks can be managed through to a stage of very consistent, safe and highly standardized care. Much surgery relies on team based approaches using the intermediate model, but some types of surgery that are very well understood can now be considered in the category of ultra-safe. Units that focus entirely on a single operation, such as cataract or hernia, achieve very consistent results and high levels of safety, though this may partly be achieved by careful patient selection. We must recognise that this approach cannot be the decision of the team alone; ultra-safe surgery requires a highly stable and controlled environment underpinned by very reliable basic processes.

In unusual circumstances any team, no matter how proceduralised the environment, may have to adapt, respond and recover. Conversely highly adaptive teams still need a core of rock solid procedures which are closely adhered to. A healthcare team might, in one afternoon, work in an ultra-safe manner at some points, such as when a care pathway is clearly defined and entirely appropriate for the patient; they may work in a high-reliability mode for the main part and, for short periods, in an ultra-adaptive mode. Longer term approaches to the underlying approach however require quite substantial adjustment not just within the team but in the wider organisation and possibly also in the regulatory environment.

A move to a new approach to safety is possible in some circumstances but it often only occurs after an event that affects the entire profession and its economy. The industrial chemical industry, for example, which in some cases is still based on adaptive models dating from the 1960s and 1970s, made a definitive switch to an HRO model after the events that occurred in Seveso in Italy in 1976 and the European Directive that followed in 1982. In this case and many others the transition to a new approach is not led from within the industry but forced by regulatory authorities and government. When this happens a prolonged period of adjustment is needed during which the system migrates gradually, loses the benefits of the previous model (a higher level of adaptation and flexibility), but gains the advantages of the new model (mainly in terms of safety). A permanent move to a new approach to safety cannot take place without changing the working conditions imposed by the external environment. A change in model must also be accepted by the front line operators and be consistent with the values and culture of both the team and the wider organisation. If you cannot change these conditions, safety improvements will be modest and restricted to local improvement. If you stay 'within the model' then one may improve safety by a factor of 10, whereas if the system can be protected and given stability then it can be moved to a different category with impressive safety gains.

Reflections on the Safety Ideal

The idea of a single ideal model of safety that applies to everything and aims to have zero accidents is too simple. Safety is only ever considered in relation to other objectives and those objectives may be valuable but also risky. We are never in the position of being able to aim for absolute safety but only to be as safe as possible given our objectives and tolerance for risk. Different contexts provoke different approaches safety, each with their own approach, advantages and limitations. The differences between these models lie in the trade-off between the benefits of adaptability and the benefits of control and safety. A different model may be intrinsically more effective, but may not be feasible in the context in question. Many aspects of healthcare for instance primarily rely on a high reliability approach but could move towards an ultra-safe model with additional resources and control of demand. However, while some change could be effected within healthcare, a more substantial adjustment would probably require a radically different approach to managing demand which is currently not politically feasible. Models of safety are ultimately context dependent and will vary by discipline, organization, governance and jurisdiction.

Key Points

- Safety is approached very differently in different environments. In some environments and professions risk is embraced, in some it is managed and in others it is controlled.
- We distinguish three classes of safety models: an ultra-adaptive approach associated with embracing risks, the high reliability approach managing risks, and the ultra-safe approach which relies heavily on avoiding risks.
- The three models reflect the degree to which the environment is unstable and unpredictable. Very high levels of safety can only be achieved in very controlled environments
- Intervention strategies must be adapted to these models, giving importance to experts in ultra-adaptive contexts, to teamwork in HRO contexts, and to standardisation, oversight and control in ultra-safe contexts.
- Healthcare has many different types of activity and clinical settings. Areas of highly standardized care, such as radiotherapy, conform to an ultra-safe model. In contrast much ward care corresponds to an intermediate model of team based care, employing a combination of standards and protocols, professional judgement and flexibility.
- Some clinical activities such as emergency surgery are necessarily more adaptive. The work may be scheduled but there is considerable hour-to-hour adaptation due to the huge variety of patients, case complexity, and unforeseen perturbations.
- All clinical areas, no matter how adaptive, rely on a bedrock of core procedures; adaptive is a relative term not an invitation to abandon all guidelines and go one's own way.
- A permanent move to a new approach to safety cannot take place without a change in working conditions imposed by the external environment. A change in model must also be accepted by front line operators and be consistent with the values and culture of both the team and the wider organisation.
- Different contexts provoke different approaches to safety, each with their own approach, advantages and limitations. The difference between models lies in the trade-off between the benefits of adaptability and the benefits of control and safety. A different model may be intrinsically safer but not be feasible in a particular context.

References

Amalberti R (2001) The paradoxes of almost totally safe transportation systems. Saf Sci 37(2):109–126

Amalberti R (2013) Navigating safety: necessary compromises and trade-offs – theory and practice. Springer, Heidelberg

Amalberti R, Deblon F (1992) Cognitive modelling of fighter aircraft process control: a step towards an intelligent on-board assistance system. Int J Man Mach Stud 36(5):639–671

Amalberti R, Auroy Y, Berwick D, Barach P (2005) Five system barriers to achieving ultra-safe health care. Ann Intern Med 142(9):756–764

Carthey J, Walker S, Deelchand V, Vincent C, Griffiths WH (2011) Breaking the rules: understanding non-compliance with policies and guidelines. BMJ 343:d5283

Grote G (2012) Safety management in different high-risk domains–all the same? Saf Sci 50(10):1983–1992

Lawton R (1998) Not working to rule: understanding procedural violations at work. Saf Sci 28(2):77–95

Morel G, Amalberti R, Chauvin C (2008) Articulating the differences between safety and resilience: the decision-making process of professional sea-fishing skippers. Hum Factors 50(1):1–16

Morel G, Amalberti R, Chauvin C (2009) How good micro/macro ergonomics may improve resilience, but not necessarily safety. Saf Sci 47(2):285–294

Reason JT (1997) Managing the risks of organizational accidents (Vol. 6). Aldershot: Ashgate.

Vincent C, Benn J, Hanna GB (2010) High reliability in health care. BMJ 340:c84

Weick KE, Sutcliffe KM (2007) Managing the unexpected: resilient performance in an age of uncertainty. Wiley, San Francisco

Wolfe T (1979) The right stuff. Random House, New York

Seeing Safety Through the Patient's Eyes

4

Patient harm happens in every healthcare setting: at home in convalescence, in the nursing home at physiotherapy, in an operating room under anaesthesia, in the hospital corridor lying alone on a stretcher, at the walk-in clinic with the paediatrician, in the emergency department awaiting physician assessment. Harm occurs as a result of failures in patient care, rather than from the natural progress of illness or infirmity. Harm may result from wrong or missed diagnosis, scheduling delay, poor hygiene, mistaken identity, unnoticed symptoms, hostile behaviour, device malfunction, confusing instructions, insensitive language and hazardous surroundings.

The trajectory of harm begins with the unexpected experience of harm arising from or associated with the provision of care, including acts of both commission and omission. The initial consequence of harm may be fleeting, temporary or permanent, including death. Harm also may not cease even when the cause is halted. The patient may experience harm during the episode of care when the failure occurred, or later, after some time has passed. Harm as it is first endured may evolve, transform and spread. Over time, untreated harm may cause further damage to the initial victim, both temporary and permanent, and to many others. (Canfield 2013)

Consider these reflections on patient harm written by Carolyn Canfield whose husband's care was very poorly managed at the end of his life. This is a description of harm written from the patient's side and in several respects it is strikingly different from the accounts of incidents and adverse events described by healthcare professionals. Three things in particular stand out:

- First, harm is conceived very broadly encompassing both serious disruption of treatment and lesser events that are more distressing than injurious.
- Second, harm for a patient includes serious failures to provide appropriate treatment as well as harm that occurs over and above the treatment provided. Very poor quality care is therefore seen as harmful and included within patient safety.
- Third, and perhaps most important, harm is seen not in terms of incidents but as a trajectory within a person's life. Both the genesis and consequences of harm occur over time and the timescales are very much longer than those normally considered. Incidents are simply those aspects of harm that are observed by healthcare professionals and, while important, they are a necessarily incomplete vision.

© The Author(s) 2016
C. Vincent, R. Amalberti, *Safer Healthcare: Strategies for the Real World*,
DOI 10.1007/978-3-319-25559-0_4

We are now moving to a rather different vision of patient safety. Our previous vision might reasonably be described as one of generally high quality healthcare punctuated by occasional safety incidents and adverse events. We now recast patient safety as the examination of serious failures and harm along the patient journey which must of course be seen in the context of the benefits of the healthcare received. This requires us to view both benefit and harm from the perspective of the patient, not because this is ideologically or politically correct, but because this is the reality we need to capture.

What Do We Mean by Harm?

Patient safety initially focused on rare, often tragic, events. However, as safety was more systematically studied it became clear that the frequency of error and harm were much greater than previously realised and that the safety of all patients needed to be addressed. Most patients are vulnerable to some degree to infections, adverse drug events, falls, and the complications of surgery and other treatments. Patients who are older, frailer or have several conditions may experience the adverse effects of over-treatment, polypharmacy and other problems such as delirium, dehydration or malnutrition. In addition, patients may also suffer harm from rare and perhaps unforeseeable events, stemming from new treatments, new equipment or rare combinations of problems that could not easily have been foreseen. If we want to assess harm from healthcare then we have to consider all these kinds of events.

Harm can be defined in various ways and there is no absolute border, particularly as the perimeter of patient safety is constantly expanding as we have discussed. Some types of events, such as a drug overdose with consequences for the patient, can be clearly described as a harmful event caused by healthcare. Harm that results from a failure of treatment is more difficult. For instance if a patient was not given appropriate prophylactic medication and then suffered a thromboembolism the harm, or at least the causation, is not so clear cut. With diagnostic delay, the notion of harm is more difficult still. Increasingly though failures to recognise deterioration and failure to institute treatment are being described as patient safety issues (Brady et al. 2013). Whether or not we would describe all these events as harmful we can all agree that they are undesirable and represent serious failures in the care of the patients concerned (Box 4.1).

Box 4.1 Examples of Types of Harm in Healthcare
General harm from healthcare
Hospital-acquired infections, falls, delirium and dehydration are examples of problems that can affect any patient with a serious illness.
Treatment-specific harm
Harm that is associated with a specific treatment or the management of a particular disease which may or may not be preventable. This would include adverse drug events, surgical complications, wrong site surgery and the adverse effects of chemotherapy.

Harm due to over-treatment
Patients may also be harmed from being given too much treatment, either through error (for instance a drug overdose) or from well-intended but excessive intervention. For example, excessive use of sedatives increases the risk of falls; people near the end of life may receive treatments that are painful and of no benefit to them.

Harm due to failure to provide appropriate treatment
Many patients fail to receive standard evidence-based care and for some this means their disease progresses more rapidly than it might. Examples include failure to provide rapid thrombolytic treatment for stroke, failure to provide treatment for myocardial infarction, and failure to give prophylactic antibiotics before surgery.

Harm resulting from delayed or inadequate diagnosis
Some harm results because the patient's illness is either not recognised or is diagnosed incorrectly. For example, a patient may be misdiagnosed by their primary care physician, who fails to refer them; the cancer advances and the outcome may be poorer.

Psychological harm and feeling unsafe
Instances of unkindness can linger in the memory of vulnerable people and affect how they approach future encounters with healthcare professionals. Awareness of unsafe care may have wider consequences if it leads to a loss of trust. For instance, people may be unwilling to have vaccinations, give blood or donate organs.

Adapted from Vincent et al. (2013)

Reflecting on the many ways in which healthcare can fail or harm patients it is clear that much of it is insidious, develops slowly and, if not addressed, may result in a crisis involving admission to hospital or other urgent treatment. A frail person in hospital who gradually becomes delirious receives care that is sub-standard to the point of being harmful but this cannot really be captured by thinking in terms of errors or incidents. When we consider care outside hospital the concept of 'an incident' breaks down even further. Consider, for instance, a patient who reacts adversely to prescribed anti-depressant medication over a period of several months culminating in an admission to hospital. We know that adverse drug reactions are implicated in about 5 % of hospital admissions (Winterstein et al. 2002; Stausberg et al. 2011) but the harm preceding these events has a timescale of months. Moreover harm of this kind cannot be seen in terms of 'error' at least not an error on any specific occasion.

Safety and Quality of Care from the Patient's Perspective

When we view our care as patients we see the course of our disease and the care we receive over time and in the context of our lives. Of course there are episodes of care but we assess our experience and the impact of healthcare in terms of the totality of

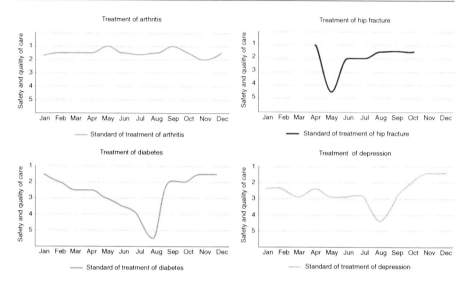

Fig. 4.1   Four patient journeys

our treatment and the overall balance of benefit and harm. How can we represent this in terms of the safety and quality of care and the various scenarios that we might wish to encompass?

Consider the four patient journeys represented in Fig. 4.1. In each case the graph provides a simple representation of the quality of care each patient receives over 1 year for the treatment of different conditions. The horizontal axis represents time and the vertical axis shows the standard of care provided (not the health of the patient or severity of their condition). The five levels discussed in Chap. 2 are shown on the left with good care shown at the top of the graphs with deteriorating or dangerous care shown by declining levels.

- The first person is receiving long-term treatment for osteoarthritis. Both treatment and monitoring are of a high standard, consistent over time and the overall quality of care is excellent.
- The second person sustains a fracture of the hip in April. Initial treatment is excellent, with prompt admission to hospital and rapid and effective surgery. However, in the post-operative period the patient develops a serious wound infection which is not immediately recognised by the nurse visiting at home. The infection worsens, a second admission to hospital is required but the infection is effectively treated and recovery is then uneventful. Overall quality of care is good, with one serious lapse during May.
- The third person is being treated for diabetes, initially effectively. However from the beginning of the year care deteriorates in that monitoring is not effective and

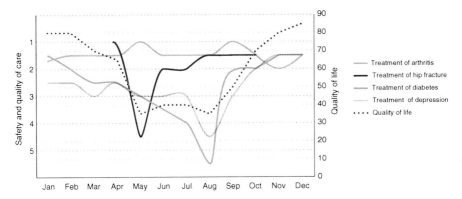

Fig. 4.2 Varying standards of care over time

becomes progressively more sporadic, resulting in admission to hospital with potentially life threatening complications in August. Hospital care is excellent however and the patient makes a good recovery.

• The fourth person is receiving partially effective treatment for depression. Monitoring is infrequent, treatment is not optimal but there are no major crises during the year. Quality of care hovers around our level 3 – poor, but still reasonably safe.

These four scenarios are relatively straightforward and, in many healthcare settings, the patient's care and progress could be monitored to some extent. We could assess the outcome of the hip replacement, record the post-operative infection and monitor the frequency of treatment for arthritis, diabetes and depression in primary care. However, we should now consider a different but more realistic scenario that more truly represents the balance of benefit and harm that we ideally wish to capture, understand and influence. This is the person who suffers from and receives treatment for a number of conditions at the same time.

Consider a person who suffers from arthritis, diabetes and depression and who sustains a hip fracture during the year. Figure 4.2 shows the four individual graphs combined for this patient with an additional second axis to give an assessment of the overall quality of life for this patient during this year of tribulation. As before the care for arthritis remains of high quality throughout the year, the care of the hip fracture generally very good but interrupted by an initially unrecognised post-operative infection. Treatment of diabetes remains problematic and we might suspect, from looking at the decline in quality of care after April, that the hospital admission and subsequent infections may have disrupted the usual monitoring. We can also very clearly see how complex healthcare is from the perspective of the patient and that very few of the healthcare professionals involved are likely to have a sense of the impact of good and poor healthcare on this person's life.

Safety Through the Patient's Eyes

In this chapter we have conceptualised safety in the context of a patient's healthcare journey showing good quality care but also encompassing a number of types of serious failure and harm. The implications of this way of approaching safety will be explored in detail in the second section of the book but it will be useful to indicate some general directions here.

The Patient Potentially Has the Most Complete Picture

The most obvious point to emerge from studying treatment over time is that the patient is, even more than in hospital, a privileged witness of events. Patient reported outcome measures are of course already a high priority, but we clearly need to begin to find ways of tracking patient experience of healthcare over time and integrating this information with available clinical information. This is easy to say but likely to be a task of considerable difficulty.

The Healthcare professional's View Is Necessarily Incomplete

Each healthcare professional involved with a patient will only have a partial view of the patient journey. Even within hospital, whether notes are electronic or paper, it can be difficult to understand the trajectory of patient care. The problem is even more acute outside hospital. A good general practitioner or family doctor is best positioned to understand the full patient journey, but we will need to develop methods of representing the full perspective of care that can be shared across different settings.

The Resources of the Patient and Family Are Critical to Safe Care

Increasingly patients and families are managing the complex work of coordinating their care. The formal assessment of these resources, financial, emotional and practical will become essential to the coordination of care and the idea of the patient as part of the healthcare team will move from being an aspiration to a necessity. This can certainly bring benefits in terms of patient engagement and patient empowerment but also carries risks as patients shoulder the burden of organising and delivering care and the locus of medical error moves from professionals to patients and families.

Coordination of Care Is a Major Safety Issue

Patients with multiple problems already have multiple professionals involved in their care and face major challenges in coordinating their own care. Poor communication across different settings is frequently implicated in studies of adverse events in hospital and in inquiries into major care failures in the community. Safety interventions in these settings may be less a matter of care bundles and more concerned with wider organisational interventions to ensure rapid response to crises and coordination between agencies.

Rethinking Patient Safety

At the beginning of this chapter we argued that the way we currently view patient safety assumes a generally high quality of healthcare punctuated by occasional safety incidents and adverse events. Increasingly we see this as a vision of safety from the perspective of healthcare professionals. This is a sincere vision in that professionals naturally assume that for the most part they are giving good care though they know that there are occasional lapses.

In contrast we have expanded our view of harm and recast patient safety as the examination of the totality of serious failures and harm within the patient journey which must necessarily be set against the benefits of the healthcare received. This is a vision of safety from the perspective of the patient, carer and family.

We believe that future progress in safety depends on conceptualizing safety in this broader manner and linking our understanding of safety with the wider ambitions and purposes of the healthcare system. This means viewing the risks and benefits of treatments over a longer timescale across different contexts and critically within the realities of a fragmented system with multiple vulnerabilities. This will require moving from a focus on specific errors and incidents to examining the origins of more fundamental failures of care such as avoidable hospitalisation due to undetected deterioration in a long term condition. The longer term aim both for patients and for patient safety is to consider how risk and harm can be minimised along the patient journey in pursuit of the optimum benefits from healthcare. In the following chapters we develop these ideas in more detail and consider how this ambitious, but we believe necessary, programme might be undertaken.

Key Points

- Patients have a different view of harm to professionals. Harm is conceived very broadly encompassing both serious disruption of treatment and lesser events that are more distressing than injurious.
- Harm for a patient includes serious failures to provide appropriate treatment as well as harm that occurs over and above the treatment provided. Both benefit and harm are seen not in terms of incidents but as a trajectory within a person's life.
- Many patient-identified events are not captured by the incident reporting system or recorded in the medical record.
- We propose that patient safety should focus on the totality of harm within the patient journey which must necessarily be set against the benefits of the healthcare received. This is a vision of safety from the perspective of the patient, carer and family.
- Patients and families will increasingly need to be actively involved in promoting safety. This can certainly bring benefits in terms of patient engagement and patient empowerment but also carries risks as patients shoulder the burden of organising and delivering care and the locus of medical error moves from professionals to patients and families.
- We need to view the risks and benefits of treatments over a longer timescale, across different contexts and within the realities of a fragmented system with multiple vulnerabilities. This will require moving from a focus on specific incidents to examining more fundamental longer term failures such as avoidable hospitalisation due to undetected deterioration in a chronic condition.
- We believe that future progress in safety depends on conceptualizing safety in this broader manner and linking our understanding of safety with the wider ambitions and purposes of the healthcare system.

References

Brady PW, Muething S, Kotagal U, Ashby M, Gallagher R, Hall D, Goodfriend M, White C, Bracke TM, DeCastro V, Geiser M, Simon J, Tucker KM, Olivea J, Conway PH, Wheeler DS (2013) Improving situation awareness to reduce unrecognized clinical deterioration and serious safety events. Pediatrics 131(1):e298–e308

Canfield C (2013) It's a matter of trust: a framework for patient harm. Unpublished manuscript.

Stausberg J, Halim A, Färber R (2011) Concordance and robustness of quality indicator sets for hospitals: an analysis of routine data. BMC Health Serv Res 11(1):106. doi:10.1186/1472-6963-11-106

Vincent C, Burnett S, Carthey J (2013) The measurement and monitoring of safety. Health Foundation, London

Winterstein AG, Sauer BC, Hepler CD, Poole C (2002) Preventable drug-related hospital admissions. Ann Pharmacother 36(7–8):1238–1248

The Consequences for Incident Analysis

5

Every high-risk industry devotes considerable time and resource to investigating and analysing accidents, incidents and near misses. Such industries employ many other methods for assessing safety but the identification and analysis of serious incidents and adverse events continues to be a critical stimulus and guide for safety improvement. Analyses of safety issues always require review of a range of information and recommendations should generally not be made on the basis of a single event. Nevertheless, an effective overall safety strategy must in part be founded on an understanding of untoward events, their frequency, severity, causes and contributory factors. In this chapter we consider how these analyses might need to be extended in the light of the arguments presented in the preceding chapters.

What Are We Trying to Learn When We Analyse Incidents?

A clinical scenario can be examined from a number of different perspectives, each of which may illuminate facets of the case. Cases have, from time immemorial, been used to educate and reflect on the nature of disease. They can also be used to illustrate the process of clinical decision making, the weighing of treatment options and, particularly when errors are discussed, the personal impact of incidents and mishaps. Incident analysis, for the purposes of improving the safety of healthcare, may encompass all of these perspectives but critically also includes reflection on the broader healthcare system.

A critical challenge for patient safety in earlier years was to develop a more thoughtful approach to both error and harm to patients. Human error is routinely blamed for accidents in the air, on the railways, in complex surgery and in healthcare generally. Immediately after an accident people make quick judgments and, all too often, blame the person most obviously associated with the disaster. The pilot of the plane, the doctor who gives the injection, the train driver who passes a red light are quickly singled out (Vincent et al. 1998). This rapid and unthinking reaction has been described by Richard Cook and David Woods as the 'first story' (Box 5.1). However while a particular action or omission may be the immediate cause of an

C. Vincent, R. Amalberti, *Safer Healthcare: Strategies for the Real World*,
DOI 10.1007/978-3-319-25559-0_5

incident, closer analysis usually reveals a series of events and departures from safe practice, each influenced by the working environment and the wider organizational context (Reason 1997; Vincent et al. 2000). The second story endeavours to capture the full richness of the event without the obscuring lens of hindsight and see it from the perspective of all those involved which should, ideally, include the perspective of the patient and family.

Box 5.1 First and Second Stories

The First Story represents how people, with knowledge of the outcome and the consequences for victims and organisations, first respond to breakdowns in systems that they depend on. This is a social and political process which generally tells us little about the factors that influenced human performance before the event.

First Stories are overly simplified accounts of the apparent cause of the undesired outcome. The hindsight bias narrows and distorts our view of practice after-the-fact. As a result, there is premature closure on the set of contributors that lead to failure.

When we start to pursue the Second Story our attention is directed to people working at the sharp end of the healthcare system and how human, organisational, technological and economic factors play out to create outcomes. We need to understand the pressures and dilemmas that drive human performance and how people and organizations actively work to overcome hazards (Adapted from Woods and Cook 2002)

We previously extended Reason's model and adapted it for use in healthcare, classifying the error producing conditions and organizational factors in a single broad framework of factors affecting clinical practice (Vincent et al. 1998; Vincent 2003). The 'seven levels of safety' framework describes the contributory factors and influences on safety under seven broad headings: patient factors, task factors, individual staff factors, team factors, working conditions, organisational factors and the wider institutional context (Table 5.1).

This gave rise to a method of incident analysis published in 2000, often referred to as ALARM, because it was produced with colleagues from the Association of Litigation and Risk Management (Vincent et al. 2000). The ALARM approach was primarily aimed at the acute medical sector. A later revision and extension in 2004, known as the 'London Protocol', has been translated into several languages and can be applied to all areas of healthcare including the acute sector, mental health, and primary care. The method of analysis is known by different names in different countries, with some continuing to use ALARM and other referring to the London protocol. We use the term ALARM/LONDON to describe the essential elements of the

Table 5.1 The ALARM/LONDON framework of contributory factors

Factor types	Examples of contributory factors
Patient factors	Complexity and seriousness of conditions
	Language and communication
	Personality and social factors
Task and technology factors	Design and clarity of tasks
	Availability and use of protocols
	Availability and accuracy of test results
	Decision-making aids
Individual (staff) factors	Attitude, knowledge and skills
	Competence
	Physical and mental health
Team factors	Verbal communication
	Written communication
	Supervision and seeking help
	Team structure (congruence, consistency, leadership)
Work environmental factors	Staffing levels and skills mix
	Workload and shift patterns
	Design, availability and maintenance of equipment
	Administrative and managerial support
	Physical environment
Organisational and management factors	Financial resources and constraints
	Organisational structure
	Policy, standards and goals
	Safety culture and priorities
Institutional context factors	Economic and regulatory context
	Wider health service environment
	Links with external organisations

previous versions, which is clumsy but avoids confusion. We also propose a new extended model, which we have christened ALARME to indicate the new European flavour that has been infused.

The approach developed by James Reason and others has been enormously fruitful and has greatly expanded our understanding of both the causes and prevention of harm. The question for us now is whether this perspective needs to be adapted or extended in the light of our previous arguments. The current model has been found to be effective in many different clinical settings but is primarily aimed at the analysis of relatively discrete events; it may need some revision if we are to also examine serious failures and harm that evolves over months or even years. We may need to broaden our approach to the investigation and analysis of incidents in a number of ways.

Essential Concepts of ALARME

The ALARM/LONDON approach set out a methodology and structured approach to reflection on the many factors that may contribute to an incident. During an investigation information is gleaned from a variety of sources: Case records, statements and other relevant documentation are reviewed and interviews are carried out with staff and ideally with the patient and family. Once the chronology of events is clear there are three main considerations: the care delivery problems identified within the chronology, the clinical context for each of them and the factors contributing to the occurrence of the care delivery problems. The key questions are: What happened? (the outcome and chronology); How did it happen? (the care delivery problems) and Why did it happen? (the contributory factors) (Vincent et al. 2000).

In the context of this book there are four new issues to be considered:

- First, we need to look at a broader class of events which impact on the patient. Some events for analysis need to be selected from the patient's point of view in addition to those identified by professionals.
- Second, we need to extend the analysis to examine an episode in the patient journey rather than a single incident. The timeframe is widened to include the whole 'event journey'. ALARME proposes an extended approach that applies the classic grid of contributory factors to each of the identified care delivery problems in the unfolding story of the 'emerging harm' considered for initial analysis
- Third, we need to pay more attention to both successes and failures of detection, anticipation and recovery. We need to consider not only problems but also success, detection and recovery and how they combine to produce the overall ratio of benefit and harm for the patient. This in turn affects the nature of the learning and the subsequent safety interventions that we might recommend
- Fourth, we potentially have to adapt both methods of analysis and recommendations to the different contexts and models of safety

Our expanded process of investigation maintains the basic approach of ALARM/LONDON and the original table of contributory factors, but extends the time frame and includes analysis and reflection on success, detection and recovery (Fig. 5.1 and Table 5.2). The changes we propose would require significant research and investment in the development of new methods but we believe this is essential if safety is to be effectively managed across clinical contexts.

Select Problems for Analysis Which Are Important to Patients

We already know that patients and families are able to reliably identify adverse events that have not been detected by professionals. Patients have been shown in a number of studies to report errors and adverse events accurately and to provide additional information not available to healthcare professionals. Many

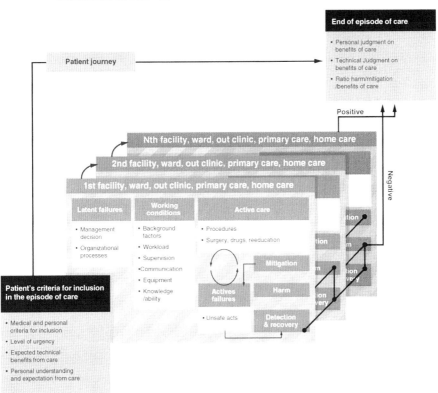

Fig. 5.1 Analysis of safety along the patient journey

patient-identified events are not captured by the hospital incident reporting system or recorded in the medical record (Weingart et al. 2005; Weissman et al. 2008). Findings from several patient surveys suggest that patients report a much higher rate of errors and adverse events of some kind than the published rates based on hospital record review (King et al. 2010; Lehmann et al. 2010). There is therefore already a case for selecting some patient identified incidents for analysis even in hospitals; outside hospitals where patients and carers probably have the most comprehensive picture of care, the argument is stronger still.

We suggest that events considered for analysis should be selected from the patient's point of view as well as by professionals. Tragedies of course deserve full and comprehensive investigation, but insights into safety may emerge from many types of event. We do not yet know what other kind of events might be identified as worthy of investigation by patients and families. What makes an event 'abnormal', and therefore a potential candidate for analysis, is a complex matter of surprise, rarity, intensity, severity and perhaps also the a basic feeling of the injustice of being

Table 5.2 New features of ALARME

ALARM/LONDON	ALARME
Identification and decision to investigate	Ask patients to tell their story of the episode of care, focusing both on what went well and poorly; select some of these cases for analysis
Select people for investigation team	Include patient and family where possible
Organisation and data gathering	Ask patients and family to tell their story and reflect on contributory factors
Determine accident chronology	Widen the timeframe to the whole patient journey
Identify Care Delivery Problems (CDPs)	Identify benefits of care as well as problems, and include detection and recovery from problems
Identify contributory factors	Identify contributory factors to each individual problem and to detection and recovery
Support for patients, families and staff not explicitly considered	Reflect and comment on disclosure process and support for patients, families and staff
Recommendations and developing an action plan	Select from the full portfolio of strategies and interventions

injured in a place of safety. The events identified by patients and families may be quite different from those identified by professionals.

Future reporting systems which seek to involve patients and families will need a balance of open-ended narratives and closed-ended questions for cause analysis and classification (King et al. 2010). Primary care patient reporting studies have used a combination of methods such as written, online or telephone reporting and telephone recruitment with a follow up in-person interview. Interviewing patients in person is particularly effective when following up hospital patients; the highest response rate overall in published studies is 96 %, achieved by in-person patient advocate interviews for a specific hospital unit. However, there are still many barriers to the use of patient derived information, particularly a lack of support for the values of patient centred care, and consequent risk of low value attached to patient involvement (Davies and Cleary 2005).

Widen the Time Frame of Analysis: Review the Patient Journey

Many serious events occur because of multiple small failures in the care of a patient rather than any single, dramatic failure. Sometimes these individual failures combine at a single time when, for instance, a young doctor is unsupervised at night with inadequate equipment, a difficult team and a very sick patient. More often though, in the care of a patient over time, we see a progressive degradation in care due to a combination of errors and system vulnerabilities and sometimes neglect. Advances in patient safety are severely hampered by the narrow timeframe used in incident detection and analysis.

We already understand that after an incident we need to look back to the series of events that led up to the problem and which are directly or indirectly linked to it.

Amalberti and colleagues (2011) have previously argued that we should extent the time frame of analysis to consider an 'event journey' (Amalberti et al. 2011). However to examine safety over longer time periods, particularly in community settings, we now believe that we should speak simply of the patient journey. This means looking back through the medical history of the patient in search for all events that have defined the patient's journey and contributed to the final outcome, whether or not these events have been perceived as serious at the time they occurred and assessing whether the problem was detected and resolved. Most important of all, the event would ideally be examined through the eyes of the patient and family as well as the eyes of the professionals.

The selection of the time frame of the analysis depends on the conditions suffered by the patient, the nature of the problems identified and the complexity of the patient journey. Standard episodes of care are easily identified; a hip replacement for instance could cover the period between the initial decision to operate and the completion of the rehabilitation process. Alternatively, depending on the nature of the safety issues identified, one might focus on a particular period such as from the original operation through to rehabilitation at home. The most important development is to begin by examining a period of care rather than a specific incident and its antecedents. Consider three different timeframes to detect and analyse events associated with the occurrence of complications. The shortest timeframe (A) would cover simple problems relating to the direct coupling between a wrong action and the immediate consequence to the patient (such as mistaken identification). A somewhat longer timeframe (B) would cover the events leading up to a medical complications and its subsequent management which might encompass an entire acute care episode from initial admission, to discharge and rehabilitation. The longest timeframe (C) might cover several months leading up to an avoidable hospital admission, the time spent in hospital and subsequent recovery. In-hospital and short-term (30- or 60-day) post-discharge mortality might be used as a starting point to investigate opportunity targets, avoidable mortality, and other indicators for complications.

The original ALARM/LONDON protocol proposed that, after the initial care delivery problems were identified, each should be analysed separately to consider the contributory factors (Vincent et al. 2000). In a sequence of problems different sets of contributory factors may be associated with each specific problem. For instance a young nurse or doctor might fail to ask for advice about a deteriorating patient due to inexperience, poor supervision and deficiencies in teamwork; in contrast the same patient might later fail to receive the correct medication, but this might be due to inadequate staffing and poor organisation of care. In practice the full analysis is seldom done and all the contributory factors are considered together as if all were relevant to the single event. However, this more subtle perspective becomes much more important with a longer timescale as a series of problems may be identified which are clearly separated in time and context. Each of these can be separately analysed using the ALARM grid to build up a much more detailed picture of system vulnerabilities.

Figure 5.2, describing the causes and response to an adverse drug event, provides an example of the new approach. The example shows the triple value of

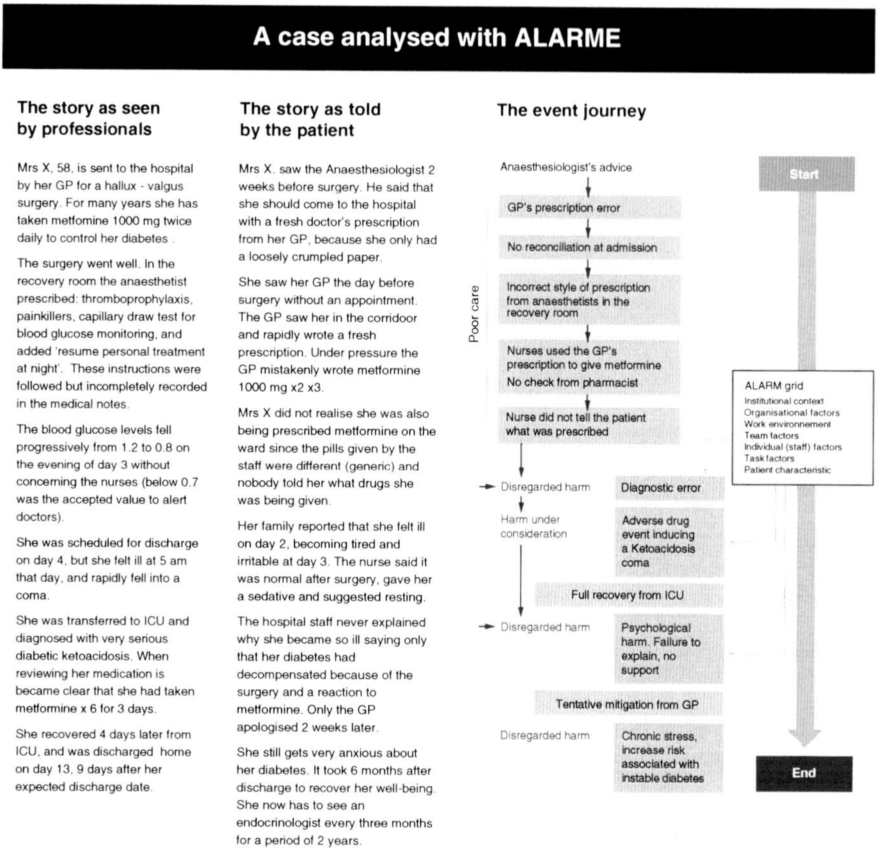

A case analysed with ALARME

The story as seen by professionals

Mrs X, 58, is sent to the hospital by her GP for a hallux - valgus surgery. For many years she has taken metfomine 1000 mg twice daily to control her diabetes .

The surgery went well. In the recovery room the anaesthetist prescribed: thromboprophylaxis, painkillers, capillary draw test for blood glucose monitoring, and added 'resume personal treatment at night'. These instructions were followed but incompletely recorded in the medical notes.

The blood glucose levels fell progressively from 1.2 to 0.8 on the evening of day 3 without concerning the nurses (below 0.7 was the accepted value to alert doctors).

She was scheduled for discharge on day 4, but she felt ill at 5 am that day, and rapidly fell into a coma.

She was transferred to ICU and diagnosed with very serious diabetic ketoacidosis. When reviewing her medication is became clear that she had taken metformine x 6 for 3 days.

She recovered 4 days later from ICU, and was discharged home on day 13, 9 days after her expected discharge date.

The story as told by the patient

Mrs X. saw the Anaesthesiologist 2 weeks before surgery. He said that she should come to the hospital with a fresh doctor's prescription from her GP, because she only had a loosely crumpled paper.

She saw her GP the day before surgery without an appointment. The GP saw her in the corridor and rapidly wrote a fresh prescription. Under pressure the GP mistakenly wrote metformine 1000 mg x2 x3.

Mrs X did not realise she was also being prescribed metformine on the ward since the pills given by the staff were different (generic) and nobody told her what drugs she was being given.

Her family reported that she felt ill on day 2, becoming tired and irritable at day 3. The nurse said it was normal after surgery, gave her a sedative and suggested resting.

The hospital staff never explained why she became so ill saying only that her diabetes had decompensated because of the surgery and a reaction to metformine. Only the GP apologised 2 weeks later.

She still gets very anxious about her diabetes. It took 6 months after discharge to recover her well-being. She now has to see an endocrinologist every three months for a period of 2 years.

The event journey

Anaesthesiologist's advice
↓
GP's prescription error
↓
No reconciliation at admission
↓
Incorrect style of prescription from anaesthetists in the recovery room
↓
Nurses used the GP's prescription to give metformine
No check from pharmacist
↓
Nurse did not tell the patient what was prescribed

Poor care

→ Disregarded harm → Diagnostic error

Harm under consideration → Adverse drug event inducing a Ketoacidosis coma

Full recovery from ICU

→ Disregarded harm → Psychological harm. Failure to explain, no support

Tentative mitigation from GP

Disregarded harm → Chronic stress, increase risk associated with instable diabetes

Start

ALARM grid
Institutional context
Organisational factors
Work environnement
Team factors
Individual (staff) factors
Task factors
Patient characteristic

End

Fig. 5.2 A case analysed with ALARME

ALARME: first gathering the story of the event journey from the patient's perspective to give a more complete account; second, widening the scope of the analysis to the full patient journey to include pre-admission and events after discharge from hospital; third identifying and analysing other usually disregarded events to reveal the cumulative impact of poor care, initial deterioration and eventual recovery.

This broader approach will require a new type of meeting (probably video conference) covering longer periods in the patient's medical history and involving the participation of both hospital and community practitioners. It would also require the development of new indicators and electronic traces, such as tools to monitor individual patients' lab results, to record the nature and duration of all breakdowns in the continuum of care. A full picture would require tracking the treatment of all the disorders from which the patient suffered not just single diseases.

Success and Failure in Detection and Recovery

In most systems errors are relatively frequent but few impact on safety because of the capacity of humans and organisations to recover from errors. In aviation, for example, numerous studies show that professional pilots make at least one clear error per hour, whatever the circumstances and the quality of the workplace design (Helmreich 2000; Amalberti 2001). The great majority of errors made are rapidly detected by the person who made them, with routine errors being better detected than mistakes. Experts of course make fewer errors overall than novices but the best marker of high level expertise is the detection of error rather than its production. Success in detection of errors is the true marker of expertise, while error production is not. Detection and recovery are sensitive to high workload, task interruptions, and system time management (Amalberti et al. 2011; Degos et al. 2009).

What are the implications for safety and for the analysis of incidents? We commonly assume that the best way to make a system safer is to reduce the number of errors and failures. This is, in many cases, entirely reasonable. Automation for instance, or reminder systems, can have a massive impact on minor errors. A more organized handover process might enhance the transfer of essential information. However eliminating all errors, which would mean considerably restricting human behaviour, is not possible and arguably not desirable.

We need in practice to distinguish errors that have immediate consequences for the patient and those which can be considered as minor deviations in the work process which can be noticed and corrected. The first class of errors do indeed need formal, rigorous rules to protect the patient, such as clear protocols for the management of electrolytes or multiple and redundant patient identification checks. For the many millions of other minor errors it is more efficient and effective to rely on detection and recovery by means of self-awareness and good coordination and communication within the team. These findings also suggest that reliable human-system interaction will be best achieved by designing interfaces that minimize the potential for control interference and support recovery from errors. In other words, the focus should be on control of the effect of errors rather than on the elimination of error per se (Rasmussen and Vicente 1989).

The standard approach for incident analysis in healthcare has primarily focused on identifying the causes and contributory factors of the event, with the idea that this will allow us to intervene to remove these problems and improve safety. These strategies make perfect sense in any system which is either highly standardised or at least reasonably well controlled, since there it is clearly possible to implement changes that address these vulnerabilities. The recommendations from many analyses of healthcare incidents are essentially recommendations to improve reliability (such as more training or more procedures) or to address the wider contributory factors such as poor communication or inadequate working conditions. In all cases we attempt, quite reasonably, to make the system more reliable and hence safer.

We could however expand the scope of the inquiry and the analysis. There is much to learn from the ability of the system to detect and recover from failures and close calls

(Wu 2011). For example, in addition to identifying failures and contributory factors we could instead ask 'what failures of recovery occurred in the care of this patient?' and 'how we can we improve detection and recovery in settings such as these?' This would have implications both for our understanding of events and, more importantly, for the recommendations which follow such analyses which might expand to include a much stronger focus on developing detection and recovery strategies.

Adapting the Analysis to Context

In addition to the developments described above we believe that we may also have to extend our thinking by adapting methods of analysis to the different contexts and models of safety we have outlined. We should be clear at this point that we do not, as yet, know how to do this. Many authors, particularly Erik Hollnagel, have drawn attention to the need for a wider array of accident models which are better adapted to fluid and dynamic environments (Hollnagel 2014). However we do not as yet have sufficient understanding to match models to environments and we have certainly not developed practical methods of analysis which are customized to different contexts.

We can however begin to consider what such an analysis might look like. Suppose we analyse an accident in a very risky unstructured environment – this might be deep sea fishing or an incident that occurred in home care involving someone with serious mental health problems. Are we looking for the same kind of causes and contributory factors as we are in a much more structured environment? The factors might be different and also the balance of factors might be different. For instance the framework of contributory factors (Vincent et al. 1998) identifies patient factors as a potential contributor to an incident. In a highly standardized environment, such as radiotherapy department, personal characteristics play a much less important role than in situations in which a person is responsible for their own care. People with serious mental health or cognitive problems are also clearly at higher risk of making drug errors in their own care. So, the relevance and influence of different types of contributory factors should be different in different contexts. This has, as far as we know, not been addressed empirically but should be entirely feasible. The next step is to ask if we should, in different contexts, be identifying different kinds of recommendations depending on the clinical context. This in turn depends on how one believes safety is achieved and realised in different settings. However before we can fully consider this issue we need to set out our proposals for a strategic approach to safety interventions addressed in the following chapters.

Key Points

- Every high risk industry devotes considerable time and resource to investigating and analysing accidents, incidents and close calls.
- Effective incident analysis requires a framework which includes guidance on the selection of incidents, and how the investigation and analysis should be conducted.
- Our current framework (known both as ALARM and London Protocol) for incident analysis in medicine: (i) identifies events for analysis chosen by professionals (ii) is based on an underlying safety model examining causes and contributory factors and (iii) uses the 'seven levels of safety' framework to guide the identification of contributory factors and potential interventions.
- The current framework remains relevant, but needs to be significantly adapted to reflect the new safety challenges.
- We need to include events that reflect harm in the eyes of patients who may identify problems that are not necessarily seen by professionals.
- We need to develop an approach which reflects the importance of poor care evolving over time, which in turn affects the nature of the learning and subsequent safety strategies that we implement.
- We propose a new approach to incident analysis (ALARME) which considers contributory factors along the whole patient journey and which includes attention to successes, failures, recovery and mitigation.
- This new approach to incident analysis involves the participation of the patient and family and both hospital and community practitioners. It may require the inclusion of new information such as the patient's personal story of illness and individual laboratory results over time.
- The changes we propose would require significant research and investment in the development of new methods but we believe this is essential if safety is to be effectively managed across clinical contexts.

References

Amalberti R (2001) The paradoxes of almost totally safe transportation systems. Saf Sci 37(2):109–126

Amalberti R, Benhamou D, Auroy Y, Degos L (2011) Adverse events in medicine: easy to count, complicated to understand, and complex to prevent. J Biomed Inform 44(3):390–394

Davies E, Cleary PD (2005) Hearing the patient's voice? Factors affecting the use of patient survey data in quality improvement. Qual Saf Health Care 14(6):428–432

Degos L, Amalberti R, Bacou J, Bruneau C, Carlet J (2009) The frontiers of patient safety: breaking the traditional mould. BMJ 338:b2585

Helmreich RL (2000) On error management: lessons from aviation. BMJ 320(7237):781

Hollnagel E (2014) Safety-I and safety–II: the past and future of safety management. Ashgate Publishing, Guildford

King A, Daniels J, Lim J, Cochrane DD, Taylor A, Ansermino JM (2010) Time to listen: a review of methods to solicit patient reports of adverse events. Qual Saf Health Care 19(2):148–157

Lehmann M, Monte K, Barach P, Kindler CH (2010) Postoperative patient complaints: a prospective interview study of 12,276 patients. J Clin Anesth 22(1):13–21

Rasmussen J, Vicente KJ (1989) Coping with human errors through system design: implications for ecological interface design. Int J Man Mach Stud 31(5):517–534

Reason J (1997) Managing the risk of organizational accidents. Ashgate, Aldershot

Vincent C (2003) Understanding and responding to adverse events. N Engl J Med 348(11): 1051–1056

Vincent C, Taylor-Adams S, Stanhope N (1998) Framework for analysing risk and safety in clinical medicine. Br Med J 316(7138):1154–1157

Vincent C, Taylor-Adams S, Chapman EJ, Hewett D, Prior S, Strange P, Tizzard A (2000) How to investigate and analyse clinical incidents: clinical risk unit and association of litigation and risk management protocol. Br Med J 320(7237):777

Weingart SN, Pagovich O, Sands DZ, Li JM, Aronson MD, Davis RB, Bates DW, Phillips RS (2005) What can hospitalized patients tell us about adverse events? Learning from patient-reported incidents. J Gen Intern Med 20(9):830–836

Weissman JS, Schneider EC, Weingart SN, Epstein AM, David-Kasdan J, Feibelmann S, Annas CL, Ridley N, Kirle L, Gatsonis C (2008) Comparing patient-reported hospital adverse events with medical record review: do patients know something that hospitals do not? Ann Intern Med 149(2):100–108

Woods DD, Cook RI (2002) Nine steps to move forward from error. Cogn Technol Work 4(2): 137–144

Wu AW (2011) The value of close calls in improving patient safety: learning how to avoid and mitigate patient harm. Joint Commission Resources, Washington

Strategies for Safety

6

Imagine that you are the leader of a healthcare unit or organisation. You are concerned about safety but you have (as always) limited time and resources. You plan a programme lasting 1 year initially and perhaps extending to 5 years.

- What should you do to improve safety?
- What safety strategies are available to you?
- How can these strategies be most effectively combined?

You might first review safety standards in your organisation and the evidence for safety improvement. From this you would probably conclude, as we have argued earlier, that there are many lapses from basic standards and that the most critical task is to improve adherence to basic safety critical procedures. This is of course easier said than done but it is the basis of most healthcare safety interventions whether this is reducing infection, improving risk assessment, avoiding wrong site surgery or improving medication safety. By this point in the book however you will have realised the near impossibility of always providing optimal care which corresponds with standards in many, if not most, settings in healthcare. Adherence to standards provides an essential foundation but not a complete vision. We may have to think a little more broadly.

What Options Do We Have for Improving Safety?

We should be wary of modelling all future safety interventions on our most visible successes. In some highly standardised areas, such as radiotherapy or management of blood products, a combination of automation and highly standardised procedures combine to deliver genuinely ultra-safe systems. However, at the other extreme, consider the care of a patient with psychosis in the community. We cannot, and should not, enforce standards and procedures in care that patients and families provide. The management of risk in such a setting clearly requires a different approach

© The Author(s) 2016
C. Vincent, R. Amalberti, *Safer Healthcare: Strategies for the Real World*,
DOI 10.1007/978-3-319-25559-0_6

based more on anticipation and detection of incipient problems and a rapid response. We have to accept and value greater autonomy and, with this greater freedom, comes greater risk. This means that safety strategies need to rely less on rules and standards and more on the detection of problems and a rapid response to them.

In the remainder of this chapter we outline the main strategies for improving safety in healthcare that can be used by our imaginary clinical leader or manager. Our hope is that providing a high level architecture of safety strategies will support frontline leaders and organisations in devising an effective safety programme. Rather than adopting piecemeal solutions we believe that we need to first articulate a high level vision of what strategies are available and how they might be employed in each setting. As we will see some strategies are most useful in highly standardised areas of work while others come to the fore in more fluid and dynamic environments. None of them in isolation necessarily provide a high level of patient safety. The aim is to find a blend of strategies and interventions appropriate to the context and the organisation.

Five Safety Strategies

We outline five broad strategies (Box 6.1) each of which is associated with a family of interventions. The strategies are, we believe, applicable at all levels of the healthcare system from the frontline to regulation and governance of the system. Two of the strategies we discuss aim to optimise the care provided to the patient. The other approaches are focused on the management of risk and the avoidance of harm.

Box 6.1 Five Safety Strategies

Safety as best practice: aspire to standards – Reducing specific harms and improving clinical processes

Improving healthcare processes and system – Intervening to support individuals and teams, improve working conditions and organisational practices

Risk control – Placing restrictions on performance, demand or working conditions

Improving capacity for monitoring, adaptation and response.

Mitigation – Planning for potential harm and recovery.

The first two strategies approaches aim, broadly speaking, to achieve safety by optimising care for the patient. In a sense safety and quality and equated; the aim is to provide care at levels 1 and 2. Within this general approach we distinguish focal safety programmes aimed at specific harms or specific clinical processes (Safety as best practice) and more general attempts to improve work systems and processes across a number of clinical settings (Improving the system). These approaches are well described in the patient safety literature and we will only briefly summarise the main features here as our primary purpose is to draw attention to other important and complementary approaches.

Optimisation of processes and systems is indeed optimal if it can be made to work. The difficulty is that in the real world optimal care is usually not achievable for at least some of the time. Once there is evidence of a substantial departure from best practice then the question becomes how best to manage those departures and the associated risk. The remaining three approaches are risk management strategies: risk control; monitoring, adaptation and response; and mitigation. Optimisation strategies improve efficiency and other aspects of quality as much as they improve safety. In contrast risk control, adaptation and recovery strategies are most concerned with improving safety.

Safety problems are also sometimes resolved because of the introduction of a completely new way of investigating or treating an illness. The development and rapid adoption of laparoscopic surgery for instance means that patients no longer have large wounds from major incisions, are less vulnerable to infections and have a much shorter hospital stay. Reduction of infection is a major safety target but was here achieved indirectly by a major surgical innovation. While we recognise that innovation often improves safety we do not consider it as a safety strategy, in the sense of a plan that can be implemented relatively quickly, because major innovations usually occur over long time periods and can only be implemented once they have been tried and tested.

Strategy I: Safety as Best Practice

The most dramatic safety improvements so far demonstrated have been those with a strong focus on a core clinical issue or a specific clinical process. They may be focussed on the reduction of a specific form of harm, such as falls or central line infections, or increasing the reliability of specific clinical processes such as pre-operative checks. We originally conceptualised this approach as 'aspiring to standards' as we regard basic standards and procedures as the foundation of safe systems, though we recognise that for an individual patient there is a great deal more to optimal care than achieving standards. In our terms 'best practice' suggests that a team or organisation aims and believes that they can provide care at levels 1 and 2 (Table 6.1)

A recent review of the patient safety literature (Shekelle et al. 2011) found only ten interventions that could be currently recommended for implementation; almost

Table 6.1 Safety as best practice: aspire to standards

Interventions	Examples
Focal safety programme: reduction of harm	Interventions to reduce central line infections
	Inpatient falls reduction programmes
	Interventions to reduce urinary catheter use and infection
	Interventions to reduce pressure ulcers
Improved reliability of targeted processes	WHO Surgical safety and other checklists
	Medication reconciliation
	Care bundles for ventilator associated pneumonia

all of them would, in our terms, be described as focal safety interventions. The essential idea is that complying with proven evidence and standards will produce optimal quality and safety. Many patients come to harm because established, scientifically based standards of practice are not reliably followed. Safety interventions of this kind first marshal the scientific evidence, then identify the core practices and endeavour to reliably bring these practices to patient care.

Box 6.2 Improving Safety by Achieving Best Clinical Practice
- Explicitly describe the theory behind the chosen intervention or provide an explicit logic model for why this patient safety practice should work
- Describe the patient safety practice in sufficient detail that it can be replicated, including the expected effect on staff roles
- Detail the implementation process, the actual effects on staff roles, and how the implementation or intervention changed over time
- Assess the effect of the patient safety practice on outcomes and possible unexpected effects, including data on costs when available
- For studies with multiple intervention sites, assess the influence of context on the effectiveness of intervention and implementation

Adapted from Shekelle et al. (2011)

It sounds simple; one identifies a standard set of safety critical procedures and supports the staff to follow them. However, in practice these are always complex, multifaceted interventions encompassing techniques, organisation and leadership (Pronovost et al. 2008). These interventions are of course far from simple and only succeed because of a sophisticated approach to clinical engagement and implementation (Box 6.2). The reduction of central line infections for instance required changes to the organisation of care, the equipment used, simplification of guidelines, engaging local multidisciplinary teams, a staff education programme, technical measurement support and a major programme of implementation.

Strategy II: Improvement of Work Processes and Systems

Accident and incident analysis and other methods reveal a great deal about the vulnerabilities in our systems and show us the range of factors which need to be addressed if we are to design a safer, high quality healthcare system. Thoughtful analyses of serious incidents reveal a range of contributory factors relating to the patient, task and technology, staff, team, working environment, organisational and institutional environmental factors (Vincent et al. 1998). This is the classic territory of the organisational accident in which immediate errors and failures are identified which are strongly influenced by wider organisational factors. These same factors

Table 6.2 Improvement of healthcare system and processes

Interventions	Examples
Individual staff	Training in key clinical processes
	Feedback on performance
Task interventions	Standardisation and simplification of processes
	Automation of key processes
	Improved design and availability of equipment
Team standardisation and specification	Structured handover
	Formalising roles and responsibilities
	Clarity of leadership and followership
	Organisation of ward care
Working conditions	Improved lighting
	Reduction of noise and disturbance
	Improved work station design
Organisational interventions	Improved levels and organisation of staffing
	Creation of new posts to improve coordination of care for patients

also point to the means of intervention and different ways of optimising the healthcare system. For instance Pascale Carayon's systems engineering approach to patient safety emphasizes interactions between people and their environment that contribute to performance, safety and health, quality of working life, and the goods or services produced (Carayon et al. 2006) (Table 6.2)

Examples of system improvements which have, amongst other objectives, had important impacts on safety include:

- The introduction of bar coding and decision support in blood collection and transfusion (Murphy et al. 2009).
- The improvement of communication and handover along the surgical pathway (de Vries et al. 2010)
- Using information technology to reduce medication errors (Bates 2000; Avery et al. 2012)
- The use of daily goals sheets to improve the reliability of ward care (Pronovost et al. 2003)

The improvement of healthcare systems is a massive topic and there are countless examples of analyses and, to a lesser extent, interventions that are within this tradition. We cannot describe these in detail and in any case they are extensively discussed elsewhere (Carayon 2011). Systems engineering, human factors and associated disciplines are not restricted to optimisation approaches in that risk control, monitoring, adaptation and recovery are sometimes considered. However we suggest that the primary drive and focus is optimisation of the healthcare system.

Strategy III: Risk Control

The next strategy and associated family of interventions is quite different form the optimisation approaches discussed above. In many industries safety is achieved by avoiding taking unnecessary risks or placing restrictions on the conditions of operation. In contrast healthcare seldom imposes limits on either professional autonomy or productivity even when safety is severely compromised. Risk control may seem to provide the answer to all risks, but avoiding risk sometimes means losing out on the potential gain that taking the risk may have allowed. Increasing risk regulation in hospitals can lead to avoidance of treating higher risk conditions, in favour of patients presenting with lower risk (McGivern and Fischer 2012). Avoidance of risk is not necessarily a good option for patients either as there are many circumstances in which clear-sightedly making a risky choice is entirely reasonable. However risk control does not aim to prevent a considered, if risky, decision but to improve the chances of a good outcome once the decision has been taken.

Risk control is widely used in other high-risk industries. The safety systems in nuclear and other facilities include numerous features which will stop the process if conditions become potentially unsafe. Commercial aviation uses a similar approach in many circumstances. For example, a storm in Miami will result in all flights to Miami grounded at their departure airport or diverted to others airports. Safety cases, the process by which potential oil and other installations are assessed, are almost unknown in healthcare. New clinical facilities are opened, or indeed closed, largely on grounds of need and cost without any formal risk assessments. Safety cases are designed to offer formal assurance and assessment that the facility can be run safely but also set out the conditions under which this can occur and building in procedures or automation to restrict activity when necessary. These are all examples of risk control by placing limits or restrictions on productive activity in the interests of safety (Table 6.3).

Table 6.3 Risk control

Interventions	Examples
Withdraw services	Close facilities if evidence of serious safety concerns
	Close facilities temporarily while safety assessments are carried out
Reduce demand	Reduce overall demand
	Reduce patient flow either temporarily or permanently
Place restrictions on services	Restrict services either temporarily or permanently
Place restrictions on individuals or conditions of operation	Define 'no-go' conditions for investigations and treatments
	Withdraw or restrict individual members of staff either temporarily or permanently
Prioritisation	Select and emphasise safety critical standards while allowing some reduction of other work, either temporarily or permanently

Healthcare does contain examples of risk control but these are seldom discussed in the context of patient safety. For instance in 2013 the medical director of the British NHS decided to temporarily close a major cardiac surgery unit because there were indications of excess mortality. He made it clear that this was just a precaution. After a few weeks, following further investigations, the unit reopened. This was seen as a very unusual intervention which caused considerable uproar. Yet, in aviation, airports are closed as soon as any substantial risk is identified.

Risk control can also be achieved by severely limiting the circumstances in which a unit can operate. Consider, for example, the way that the Australian healthcare system places very strict limits of what some clinics are allowed to do (New South Wales Government Private Facilities 2007). Some clinics are only authorised to work in a very specific medical area, and are staffed and equipped accordingly. Within this area, the facility must accept patients and deliver safe care; outside this area, the facility is not allowed to deliver care and must transfer all patients to a competent facility.

Risk control can include withdrawing services or even closing facilities when they have become dangerous. However the main thrust of this approach is to restrict the conditions in which investigations or treatment can be given. There are, for instance, very strict regulations governing the provision of radiotherapy but almost no restrictions on the conditions in which a surgical operation can go ahead. We believe that much more consideration should be given to the control of risk to protect both patients and staff from engaging in unnecessarily risky activities.

Strategy IV: Monitoring, Adaptation and Response

Safety is achieved partly by attempting to reduce errors but also by actively managing the problems and deviations that inevitably occur. Once we accept that errors and failures occur frequently in any system then we see the need to develop methods of monitoring, adapting and responding and recovering from failure. Adapting and responding to problems happens all the time in healthcare and is as relevant to managers as to frontline staff. Managers in particularly are constantly 'firefighting' and resolving problems, but this tends to be done on an ad hoc individual basis. The question we address here is whether these often improvised adaptations can evolve to become formal safety strategies in the sense of actively building such capacity into healthcare systems. Ideally senior clinicians and managers would maintain safety at a good level by playing on a palette of known and practiced organisational and cultural adjustments.

Adapting and responding is much more important in deep sea fishing than on an assembly line but all work requires this capacity to some degree. Being on the lookout for problems, adapting and working around difficulties is part and parcel of all jobs. In high risk industries such as healthcare the pattern is the same but the stakes are much higher and the capacity for rapid response and recovery may literally be a matter of life and death. This family of interventions is paradoxically the most used in daily work in healthcare but not properly developed as a strategic reality in patient safety.

The broad capacity of adapting and responding has been discussed extensively in the safety literature and made the cornerstone of some approaches to safety such as resilience engineering (Hollnagel et al. 2007). The term resilience is used in very different ways (Macrae 2014), sometimes very broadly in an attempt to describe and articulate the qualities of a safe organisation and sometimes in a more restricted sense of a capacity to adapt and recover from extreme or unusual circumstances. We believe that resilience is an important concept that needs serious consideration and further research and exploration in practice. However to avoid potential confusion we use the more everyday terms of monitoring, adaptation and recovery to denote occasions where or hazards or failures have been detected and are being actively managed or corrected.

We will describe a number of interventions associated with this approach in the following chapters and will just give some brief examples here (Table 6.4). An emphasis on the open discussion of error and system failures by senior leaders is enormously important in fostering a willingness to speak up and intervene if a patient is at risk. Clinical teams use many adaptive mechanisms, both formal and informal, to manage safety on a day to day basis. Anaesthetists for instance have a standard repertoire of prepared emergency routines which are called upon in certain situations. These routines are only seldom used and are deliberately honed and standardised so that they can be adhered to at times of considerable stress. At an organisational level we could see preparations for a possible infection outbreak in a similar way (Zingg et al. 2015). Briefings and debriefings can be used by ward staff, operating theatre teams and healthcare managers to monitor day to day threats to safety. For example, briefings carried out by operating theatre teams provide an opportunity to identify and resolve equipment, staffing, theatre list order issues before a case starts. Debriefings carried out at the end of the theatre list support reflective learning on what went well and what could be done better tomorrow.

Table 6.4 Improve capacity for monitoring, adaptation and response

Interventions	Examples
Improve safety culture	Patient and family engagement
	Culture of openness about error and failure
Monitoring, adaptation and response in clinical teams	Rapid response to deterioration
	Develop emergency response systems and routines
	Develop team cross checking and safety monitoring
	Building briefing and anticipation into clinical routines
Improve management of organisational pressures and priorities	Develop methods of predicting times of staff shortage and other pressures
	Improve managerial capacity to deal with dangerous situations
Regulatory compromises and adaptation	Negotiate time to move to new standards
	Actively manage safety during time of transition

Increasingly, briefings and debriefings are being introduced in other healthcare domains such as safeguarding adults and mental health teams (Vincent et al. 2013).

Strategy V: Mitigation

Mitigation is the action of reducing the severity, seriousness, or painfulness of some event. This strategy accepts that patients and indeed staff will sometimes be seriously affected or harmed during their healthcare and, critically, that the organisation concerned then has a responsibility to mitigate that harm. In particular we believe that organisations need to have effective systems in place to support patients, carers and staff in the aftermath of serious failures and harm. This is perhaps the most neglected aspect of patient safety (Table 6.5).

Accepting risk in healthcare seems at first glance to either be an admission of defeat or a cynical disregard for patients. However this strategy is rather more subtle and more important than one might think both at the clinical and organisational level. Planning for these occasions can seem to indicate a resigned acceptance of harm; in reality planning for recovery is humanitarian and necessary. A complete approach to safety must include the mitigation of harm, although managing complaints and litigation should not dominate attempts to improve safety.

Organisations of all kinds must insure against risk, deal with complaints and litigation, and manage the media and regulatory response Organisations must also make insurance arrangements for compensation to injured patients. At a national level countries develop medical insurance systems, such as no-fault compensation, to support patients who have been harmed. In most countries, the challenge of addressing error in medicine demands a thorough reconsideration of the legal mechanisms currently used to deal with error and harm in health care.

The basic needs of injured patients have been understood for 20 years (Vincent et al. 1994). We would all, in varying degrees, like an apology, an explanation, to know that steps had been taken to prevent recurrence and potentially financial and

Table 6.5 Mitigation

Interventions	Examples
Support for patients and carers	Rapid response and clear communication
	Designated follow up, psychological and physical support
	Plan clinical services for response to known complications
Support for staff	Peer to peer support programmes
	Temporary relief from clinical duties
	Provision of longer term support
Financial, legal and media response	Insurance of organisation
	Legal protection for organisation against unwarranted claims
	Capacity for rapid and proactive media response

practical assistance. Many patients experience errors during their treatment, whether they realise it or not, and some are harmed by healthcare. The harm may be minor, involving only inconvenience or discomfort, but can involve serious disability or death. Almost all bad outcomes will have some psychological consequences for both patients and staff, ranging from minor worry and distress through to depression and even despair. The experiences of these people tend not to be fully appreciated, and yet understanding the impact of such injuries is a prerequisite of providing useful and effective help (Vincent 2010). Healthcare organisations generally have extremely limited services to support either patients or staff in the aftermath of adverse events.

We know too that staff suffer a variety of consequences from being the 'second victim' as Albert Wu eloquently expressed it, not implying that the experiences of staff were necessarily comparable to those of injured patients (Wu 2000). We should also consider that a member of staff who has been seriously affected may well be performing poorly and be a risk to future patients; this again is rarely addressed. There are a few pioneering examples of programmes of support for both patients and staff (Van Pelt 2008) but this is an area of safety management which needs substantial development. We consider that this should be a core safety strategy; planning for recovery should include this core humanitarian element as well as managing risk and reputation.

Innovation

Safety problems are sometimes resolved because of the introduction of a completely new way of investigating or treating an illness or a new way of providing and organising care. Innovation in healthcare can take many forms, ranging from drug therapies, surgical procedures, devices, and tests, through to new forms of health professional training, patient education, management, financing and service delivery models. These innovations generally aim to provide better or more efficient care for patients, but safety may also be improved as a virtuous side effect of the action

Healthcare is remarkable in the breadth and pace of innovation which both improves safety and, as we have argued earlier, also changes the boundaries of what is acceptable and so creates new safety problems. The pace of innovation is such that medical knowledge dates extremely rapidly. In 2007, the median time before medical knowledge needed significant updating was only 5 years; 23 % of systematic reviews needed updating within 2 years and 15 % within 1 year (Shojania et al. 2007). According to both the United Kingdom National Institute of Clinical Excellence and the American Heart Association most recommendations and treatment guidelines need substantial adjustment every 5 years (Alderson et al. 2014; Neuman et al. 2014).

Many innovations in diagnosis and treatment have had a major positive impact on safety. For example safety in anaesthesia has improved about ten-fold over the last past 20 years, with the consensus being that the greatest safety gains have arisen from the introduction of new drugs and techniques for monitoring and for regional and ambulatory anaesthesia (Lanier 2006). The rapid introduction and spread of laparoscopic surgery has reduced length of stay, led to more rapid recovery and reduced risk of infection and other problems (Shabanzadeh and Sørensen 2012).

Safety may also be improved indirectly through the reorganisation of the healthcare system, particularly through realignment to a more patient centred vision. Many failures of care in community settings are due to failures of coordination and communication between agencies and across different parts of the system. One aspect of the 'burden of treatment' (Mair and May 2014) experienced by patients and families is that they have to organise and coordinate their own care to compensate for the failures of the healthcare system. If we succeed in developing more integrated systems of care across settings and populations these problems should reduce. Patients will be safer and experience fewer failures although the changes are not specifically targeted at safety.

Innovation is not really a safety strategy, although safety interventions may be innovative. New treatments or technologies are usually targeted at wider benefits for patients with reduction of risk being a secondary benefit. More importantly in this context 'innovation' cannot be deployed as a strategy in the same way as optimisation, control and recovery. Our imaginary chief executive with a 3–5 year time horizon cannot rely on innovation to solve safety problems but must nevertheless to alert to new developments that may change the nature of the problems that they are facing.

To sum up, innovation is a good example of a double-edged tool for safety. On the one hand, it is a critical determinant and means (perhaps the most significant) of improving safety in the long term. Innovation may also introduces new risks as well as resolving old ones particularly in the short term during the period of transition and disruption (Dixon-Woods et al. 2011). Safety may be degraded in the short term due to rapid diffusion of insufficiently tested new methods and uncontrolled individual experimentation.

Selection and Customisation of Strategies to Clinical Context

We hope that the delineation of the five strategies and their associated interventions is useful as a way of thinking through the approaches that might be taken to manage risk in any particular healthcare environment. These very broad strategies are seldom explicitly distinguished and some safety programmes unwittingly combine several types of strategy with somewhat different objectives. We believe that many situations do require a combination of different approaches but that it needs to be clear how and why each strategy is deployed (Fig. 6.1). We also need to consider how these strategies and associated interventions might be combined and in what proportions. Each clinical environment brings its own challenges and requires a different combination. We have set out three broad models of clinical work to illustrate this and others may need to be articulated. The management of risk in the community for instance in highly distributed health and social care systems may require a different kind of approach.

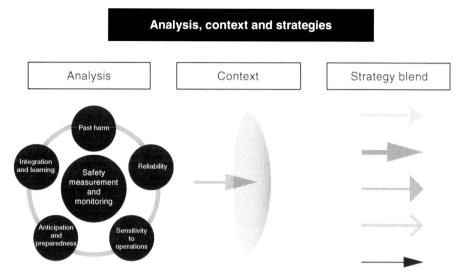

Fig. 6.1   Analysis, context and strategies

The following chapters begin to explore these ideas in more detail by providing examples of safety strategies in different settings. We can however illustrate the general idea which is that after an initial stage of diagnosis of the safety problem, illustrated by our approach to safety measurement and monitoring, the 'lens' of the clinical context will inform the particular strategy blend (see Fig. 6.1). In settings where care can be precisely defined and delineated strategies to control exposure to risk and maintain standards will predominate, hopefully accompanied by concurrent strategies to improve working conditions and support staff. In contrast in more fluid and dynamic environment strategies to improve monitoring and adaptation may be more to the fore, although all environments require a solid procedural underpinning. The next three chapters develop and illustrate these ideas in the context of hospital, home and primary care.

Key Points

- We outline a portfolio of five strategies for improving safety in healthcare each associated with a family of interventions:
  - Safety as best practice: Improving clinical processes and standards
  - Improving healthcare processes and system: Intervening to support individuals and teams, improve working conditions and organisational practices
  - Risk control: placing restrictions on performance, demand or working conditions
  - Improving capacity for monitoring, adaptation and response.
  - Mitigation: Planning for potential harm and recovery.
- Safety problems are sometimes resolved because of the introduction of a completely new way of investigating or treating an illness. These innovations generally aim to provide better or more efficient care for patients, but safety may be improved as a virtuous side effect of the innovation.
- In settings where care can be precisely delineated strategies to control exposure to risk and maintain standards will predominate. In contrast in more fluid and dynamic environment strategies to improve monitoring adaptation may be more to the fore, although all environments require a solid procedural underpinning.
- Each clinical environment brings its own challenges and requires a different combination of strategies. The management of risk in the community for instance in highly distributed health and social care systems may require a different kind of approach.

References

Alderson LJ, Alderson P, Tan T (2014) Median life span of a cohort of National Institute for Health and Care Excellence clinical guidelines was about 60 months. J Clin Epidemiol 67(1):52–55

Avery AJ, Rodgers S, Cantrill JA, Armstrong S, Cresswell K, Eden M, Elliott RA, Howard R, Kendrick D, Morris CJ, Prescott RJ, Swanwick G, Franklin M, Putman K, Boyd M, Sheikh A (2012) A pharmacist-led information technology intervention for medication errors (PINCER): a multicentre, cluster randomised, controlled trial and cost-effectiveness analysis. Lancet 379(9823):1310–1319

Bates DW (2000) Using information technology to reduce rates of medication errors in hospitals. Br Med J 320(7237):788

Carayon P (ed) (2011) Handbook of human factors and ergonomics in health care and patient safety. CRC Press, New York

Carayon P, Hundt AS, Karsh BT, Gurses AP, Alvarado CJ, Smith M, Brennan PF (2006) Work system design for patient safety: the SEIPS model. Qual Saf Health Care 15(suppl 1):i50–i58

de Vries EN, Prins HA, Crolla RM, den Outer AJ, van Andel G, van Helden SH, Schlack WS, van Putten MA, Gouma DJ, Dijkgraaf MGW, Smorenburg SM, Boermeester MA (2010) Effect of a comprehensive surgical safety system on patient outcomes. N Engl J Med 363(20):1928–1937

Dixon-Woods M, Amalberti R, Goodman S, Bergman B, Glasziou P (2011) Problems and promises of innovation: why healthcare needs to rethink its love/hate relationship with the new. BMJ Qual Saf 20(suppl):i47–i51

Hollnagel E, Woods DD, Leveson N (eds) (2007) Resilience engineering: concepts and precepts. Ashgate Publishing, Guildford

Lanier WL (2006) A three-decade perspective on anesthesia safety. Am Surg 72(11):985–989

Macrae C (2014) Close calls: managing risk and resilience in airline flight safety. Palgrave Macmillan, London

Mair FS, May CR (2014) Thinking about the burden of treatment. BMC Health Serv Res 14:281

McGivern G, Fischer MD (2012) Reactivity and reactions to regulatory transparency in medicine, psychotherapy and counselling. Soc Sci Med 74(3):289–296

Murphy MF, Staves J, Davies A, Fraser E, Parker R, Cripps B, Kay J, Vincent C (2009) How do we approach a major change program using the example of the development, evaluation, and implementation of an electronic transfusion management system? Transfusion 49(5): 829–837

Neuman MD, Goldstein JN, Cirullo MA, Schwartz JS (2014) Durability of class I American College of Cardiology and American Heart Association clinical practice guideline recommendations. JAMA 311(20):2092–2100

New South Wales Government Private Facilities Act (2007) http://www.health.nsw.gov.au/hospitals/privatehealth/pages/default.aspx. Accessed 2 Aug 2015

Pronovost P, Berenholtz S, Dorman T, Lipsett PA, Simmonds T, Haraden C (2003) Improving communication in the ICU using daily goals. J Crit Care 18(2):71–75

Pronovost PJ, Berenholtz SM, Needham DM (2008) Translating evidence into practice: a model for large scale knowledge translation. BMJ 337:a1714

Shabanzadeh DM, Sørensen LT (2012) Laparoscopic surgery compared with open surgery decreases surgical site infection in obese patients: a systematic review and meta-analysis. Ann Surg 256(6):934–945

Shekelle PG, Pronovost PJ, Wachter RM, Taylor SL, Dy SM, Foy R, Hempel S, McDonald KM, Ovretveit J, Rubenstein LV, Adams AS, Angood PB, Bates DW, Bickman L, Carayon P, Donaldson L, Duan N, Farley DO, Greenhalgh T, Haughom J, Lake ET, Lilford R, Lohr KN, Meyer GS, Miller MR, Neuhauser DV, Ryan G, Saint S, Shojania KG, Shortell SM, Stevens DP, Walshe K (2011) Advancing the science of patient safety. Ann Intern Med 154(10):693–696

Shojania KG, Sampson M, Ansari MT, Ji J, Doucette S, Moher D (2007) How quickly do systematic reviews go out of date? A survival analysis. Ann Intern Med 147(4):224–233

Van Pelt F (2008) Peer support: healthcare professionals supporting each other after adverse medical events. Qual Saf Health Care 17(4):249–252

Vincent C (2010) Patient safety, 2nd edn. Wiley Blackwell, Oxford

Vincent C, Phillips A, Young M (1994) Why do people sue doctors? A study of patients and relatives taking legal action. Lancet 343(8913):1609–1613

Vincent C, Taylor-Adams S, Stanhope N (1998) Framework for analysing risk and safety in clinical medicine. Br Med J 316(7138):1154–1157

Vincent C, Burnett S, Carthey J (2013) The measurement and monitoring of safety. The Health Foundation, London

Wu A (2000) Medical error: the second victim. Br Med J 320:726–727

Zingg W, Holmes A, Dettenkofer M, Goetting T, Secci F, Clack L, Allegranzi B, Magiorakos AP, Pittet D (2015) Hospital organisation, management, and structure for prevention of health-care-associated infection: a systematic review and expert consensus. Lancet Infect Dis 15:212–224

Safety Strategies in Hospitals

7

We have developed a series of ideas and proposals in the book which together laid the foundations for five safety strategies described in Chap. 6. We believe that thinking of safety strategies in this way has three major advantages: first, we can enlarge the range of safety strategies and interventions available to us; secondly we can customise the blend of strategies to different contexts and third the high level architecture of safety strategies may help us think more strategically about safety both day to day and on a long term basis

In this chapter we begin the process of exploring how these strategies might support safety in the hospital. The following chapters address home care and primary care. In each case we provide a short introduction to relevant aspects of safety in each context but do not dwell on well-established findings. Our primary purpose is to provide examples of interventions associated with each of the five strategies and to give a sense of the potential value of such an approach. We recognise that, in the longer term, considerable further empirical work would be needed to develop and confirm (or discount) our proposals.

A Little History

Hospital care has been the main focus of patient safety for two decades now and we can distinguish a series of phases of exploration and intervention. Each phase brought some success but simultaneously revealed barriers and limitations, which in turn stimulated a new phase of work in an evolving trial and error strategy. With experience and maturity, we understand more today about what is achievable and what has proved illusory. We are much more aware of how difficult it is to improve safety in both the short and long term.

What has been done in past decades? In the past 15 years we can distinguish three main phases each associated with different types of action and intervention. The earlier strategies have continued as the new ones emerged so that we now have 'a safety layer cake' of practices and interventions.

© The Author(s) 2016
C. Vincent, R. Amalberti, *Safer Healthcare: Strategies for the Real World*,
DOI 10.1007/978-3-319-25559-0_7

The Enthusiasm of the Early Years, 1995–2002

Systematic work on patient safety began in the mid 1990s with an emerging demarcation between a broad concern with quality and a specific focus on harm. In Britain for instance the development of clinical risk management, initially targeted at the reduction of litigation, brought a new emphasis on the analysis and reduction of harmful incidents and events (Vincent 1995). The methods and assumptions however remained rooted in those of quality improvement; the aim was to identify and count errors and incidents and then find ways of preventing them. Establishing reporting systems to detect and record incidents was at the core of the strategy. This approach was rapidly reconsidered as a result of both massive under-reporting, especially from doctors, and a gradual realisation of the impossibility of resolving the growing number of problems identified in reporting systems (Stanhope et al. 1999). A wider vision was needed which was provided by systemic concepts and tools imported from industry.

The Advent of Professionalism 2002–2005

In the late 1990s, James Reason provided an inspirational vision for healthcare that provided a clear demarcation between traditional approaches to quality improvement and the specific problems that arise when addressing safety (Reason 1997; Reason et al. 2001). Safety researchers, clinicians and managers took the concepts, techniques and methods from industrial safety and applied them to healthcare. These included a stronger emphasis on the role of latent organisational conditions which led to the development of methods of incident analysis derived this model (Vincent et al. 1998, 2000). Increasing attention was also given to human factors and ergonomics, following the success in improving interface and equipment design in industry, the use of information technology and a scientific approach to working conditions, stress and fatigue management (Bates 2000; Sexton et al. 2000; Carayon 2006). Accreditation and certification built on this new knowledge in requiring hospitals to establish risk management programmes and new patient safety indicators. Safety and risk management acquired a much higher profile and many new initiatives were developed across the healthcare system, but the impact on the safety of patients remained uncertain (Pronovost et al. 2006; Wachter 2010). The lack of clinical engagement was a major concern with patient safety remaining the province of enthusiasts and specialists – a curious situation given that safety, considered in terms of personal accountability, is perhaps the dominant concern of clinicians in their day-to-day work with patients.

Safety Culture, Multifaceted Interventions, and Teamwork 2005–2011

Surveys of safety culture demonstrated unequivocally that in many hospitals and other healthcare settings safety attitudes and values were far from ideal. Findings from many studies suggested an excessive blame culture, pressure on performance to the detriment of safety, little transparency towards patients and variable levels of supervision and teamwork. There was also huge variability between hospitals, within clinical disciplines and across different settings (Tsai et al. 2013). Whereas safety culture was initially seen as potentially directly impacting on safety, there was now a growing awareness that it might provide only a necessary foundation (Flin et al. 2006; Vincent et al. 2010)

However, as we have discussed, evidence began to emerge of marked improvements in specific safety problems at a local level and of the potential of wider application of approaches such as checklists, care bundles and so on (Haynes et al. 2009; Shekelle et al. 2011). Those proven safety wins on the frontline encouraged the healthcare community to believe that safety would progressively improve as more interventions were put into place. Improving safety across organisations and populations however has proved a great deal more challenging. The major difference between current views and what was imagined in the mid 2000s is that safety wins and rewards are now expected in the middle to long term rather than in the very short term.

Reflections on Safety in Hospitals

We provide this brief overview primarily to highlight the fact that approaches to safety in hospitals have primarily been optimising approaches of one kind and another, although comparatively little attention has been given to optimising the system overall as opposed to improving specific practices. Accreditation and regulations of the system might be thought to be examples of risk control and there are certainly examples of standards being set in order to minimise or avoid risks of certain kinds. However we suggest that most accreditation and regulation is essentially aimed at assessing compliance or failure to comply with defined standards of care. Regulators are sometimes forced to acknowledge that standards cannot be met and that adaptations must be made but we suggest that the dominant vision of how safety is achieved is one of adherence to standards.

Safety in Hospital: Distinguishing Current and Future Strategies

We propose that thinking in terms of an overall blend of high level safety strategies customised to different contexts will be an efficient and effective approach both to managing safety on a day to day basis and to improving safety over the long term. However before we start to illustrate how the five different strategies might be employed in hospital we need to consider a critical issue, which is that staff and organisations often have to employ a particular strategy not because of the needs of that clinical environment but to compensate for other underlying problems in the system. For instance, services such as acute medicine rely very heavily on monitoring, adaptation and recovery to observe, correct and recover from the inevitable departures from best practice and unforeseen problems that arise. However the fact that a strategy is extensively used does not necessarily mean that it is desirable; it might in fact be overused to compensate for other deficiencies such as poor reliability or inadequate staffing (Box 7.1). We therefore need to make a distinction at this point between:

- The blend of strategies currently used by an organisation
- The blend that might be desirable
- The strategies that might need to be developed or enhanced

Box 7.1. Adaptation and Compromise on the Wards
Recently while on call at the weekend I found my team looking dispirited, ploughing through 27 pages of printed jobs that were required for patients based on ten wards. There was no way these could all be done by two junior doctors. They were doing what any sensible person would do and "working round" an impossible task, rationing what was essential or urgent and what could be omitted.

A large proportion of the workload is phlebotomy, taking bloods and chasing the results. These should be taken by technicians but they have a fixed contract for 4 h meaning that they only deal with a small proportion of the overall workload. Tests are ordered by weekday teams, and handed over to the weekend team to check results, often without a clear indication of the purpose of the tests or what to do with the results. The weekend teams only become aware when blood has not been taken when they check for the result, leading to considerable delay in monitoring patients. There is huge variability in the clarity of the requests, the background information given, the appropriateness of the job itself and what to do with results, all compounded by the inexperience and insecurity of junior doctors on call at weekends.

Inada Kim (personal communication 2015)

Staff in all environments rely on workarounds such as obtaining information from patients rather than their health records, or using disposable gloves as tourniquets. In some cases, risks are taken such as making clinical decisions without information, or transferring used sharps to sharps bins in remote locations (Burnett et al. 2011). Often front-line coping and adaptation leads to short-term "fixes" that put off more fundamental, long-term solutions. These clinical work-arounds may also allow managers to protect themselves from inconvenient truths and shift accountability for failure to front-line workers (Wears and Vincent 2013).

We therefore always need to think, when formulating the overall approach to safety, both about what the approach is now and what might be the most effective strategy in the longer term. We certainly believe that adaptive strategies should be further developed in the sense of being planned and to some extent formalised. However this is very different from the current reliance on ad hoc improvising to compensate for missing information, faulty equipment and the like. Figure 7.1

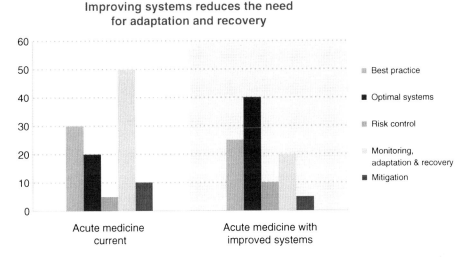

Fig. 7.1 Improving systems reduces the need for adaptation

illustrates these ideas in the context of acute medicine suggesting that increasing reliability and controlling flow and demand would reduce the need for adaptation and improvisation. With this in mind we now illustrate the five safety strategies in the context of the hospital; we devote most space to risk control, adaptation and mitigation as the other two strategies are already well described.

Safety as Best Practice

We have already given a number of examples of optimising strategies in hospitals in Chap. 6 and earlier in the book. Clearly one needs to consolidate and develop the approaches that aim to improve adherence to best clinical practice and thereby make care safer for patients. Reduction of pressure ulcers, reduction of catheter-associated infections, improved hand hygiene, improved patient identification and so on are obviously critical. Such standardised tasks and processes can be routinely audited to ensure that standards are maintained. All hospital environments, no matter how fluid and dynamic the workflow, have many core basic procedures which need to be followed. Programmes to improve adherence to basis procedures are always an important foundation for safety though never a complete solution.

Improving the System

Although the broad field of human factors and ergonomics is both huge and critically important in hospital settings we will not consider it in detail here. This is because it has been extensively discussed elsewhere and accepted as a valid and essential approach to improving safety as well as effectiveness and experience. Under this broad heading we would include improvements to the administration of medication in terms of standardizing formularies and protocols, the introduction of information technology in all its forms, formalising roles and responsibilities in clinical teams, the use of care bundles and daily goals to organise ward care and all efforts to improve basic working conditions. Improving safety through best practice and raising standards tends to require additional effort from frontline staff, at least in the early stages. There is an equal need to give attention to improving the system in order to reduce the burden on staff and so allowing more time for safety monitoring and improvement. Improvement of working conditions could involve improvements to interface design, to the ergonomics of equipment, the physical working environment or the reduction of interruptions and distractions that greatly increase propensity to error. We provide one example to illustrate the potential of this kind of approach.

Reducing the Burden on Staff: Simplification and Decluttering

The improvement of the working lives of staff is a central aim of human factors work. If we want staff to spend time monitoring and improving safety we have to create time for it and not rely on enthusiasts working at weekends and in the evenings. This means that less time must be allocated to something else and decisions must be made about what can be stripped out of the current work process. We will briefly consider the issue of policies and procedures in the British NHS as an example of how we might begin to simplify the system and reduce the burden on staff.

Within the National Health Service (NHS), a vast number of policies and guidelines govern all aspects of the work of the organisation. In an analysis of clinical guidelines related to frontline care Carthey and colleagues (2011) found that in the first 24 h of a patient admitted for emergency surgery on a fractured neck of femur there were 76 applicable guidelines. A brief survey of 15 NHS hospitals in England who published their policies on their websites showed that they had between 133 and 495 policies covering everything from dress code to medication dispensing. The average policy was 27 pages long with length varying between 2 and 122 pages (Fig. 7.2). An average hospital has 8000 pages of policies on their websites running to more than two million words (Green et al. 2015).

The plethora of unusable quasi-legal policies is an unconscionable burden on the staff, a drain on resources and paradoxically a threat to safety. First, safety critical essential procedures and other trivial policies are not sufficiently distinguished and all formal policies become degraded. Second, staff cannot possibly comply with even a fraction of the guidelines and procedures they have to contend with. Third, huge amounts of time and resource are devoted to producing policies which are more or less unusable in practice and distract from other potentially more fruitful approaches to safety. How many such procedures can we reasonably put in place in

Guidelines for fractured neck of femur in the first 24 h

Guidelines for investigations while patient is In A&E
3. Standard ECG for elderly
4. Urine dipstick
5. Blood tests
6. Chest X-ray
7. Hip X-ray
8. MRSA screen

Guidelnes to monitor and manage patients while In A&E
9. Vital signs (full set)
10. Pain score
11. Analgesia prescribed (analgesic Ladder)
12. DVT prophylaxis
13. Oxygen administation
14. IV fluids
15. Keep nil by mouth until definitive plan made
16. Transfer to orthopaedic ward within 4 h of arrival

Transfer to ward
29. Transfer guidelines
30. Patient handover

Pre-operative preparation guidelines
45. Surgery within 48h and during day time
46. Pre-operative assessment
47. Pre-operative fasting
48. Drug administration
49. Pre-operative nursing preparation
50. Pre-operative checks and accompanying a patient to theatres
51. Antibiotics prophylaxis prescription (intra + post-operative)
52. Consent + operation site mark

Intra-operative care guidelines
53. Theatre arrival checklist
54. Anaesthetic care (multiple)
55. Surgical safety checklist
56. Surgical operation (multiple)
57. Scrub nurse guidelines (multiple)
58. Radiation exposure
59. Sterility + laminar flow
60. Additional guidelines depending on circumstances eg blood transfusion

Post-operative care guidelines
61. Immediate care of patient in theatre recovery (multiple)
62. Post-operative infection prevention
63. Post-operative monitoring and investigations
64. Wound care management
65. Post-operative nutrition & supplementation
66. Post-operative surgical care, eg timing of drain removal etc.
67. Post-operative analgesia

Hospital admission
1. Procedure for triage assessment of patient arriving in A&E
2. Patient identification

Specific clinical guidelines
17. Does patient satisfy fast track criteria (elderly)
18. Full history and examination of every organ system
19. Assessment of social circumstances
20. Exclude & treat other injuries
21. Patient's orientation
22. Assessment for multiple pathology
23. Assessment of injury
24. Consider possibility of elderly abuse
25. If SpO_2 <94 % check ABG and administer oxygen
26. Treat cardiac arrhythmias according to guidelines
27. Consider and treat community acquired pneumonia
28. Consider need for bone protection medication

Ward admision/nursing checks
31. Ward orientation, information leaflet for patient and relatives
32. Assess skin and pressure areas
33. Moving and handling guidelines
34. Nutrition assessment
35. Guidelines for clinical observations, vital signs, weight and height.
36. Continence assessment
37. Information documentation

Guidelines for managing patient care
38. Discharge planning
39. Pain management
40. Investigations
41. Drug administration
42. Patient positioning, traction, immobilization and manual handling
43. Pressure area care
44. Bone protection medication guideline for elderly care

68. Rehabilitation guidelines
69. Multidisciplinary assessment for rehabilitation
70. Early post-operative mobilisation (within 24h)
71. Slips, trips and falls
72. Guidelines for exercise regime and rehabilitation

Discharge planning
73. Safe discharge and follow up
74. Involve social services if appropriate
75. Bone health assessment and treatment at discharge

Patient arrives in A&E following injury

Patient admitted to orthopaedic ward

Peri-operative care of patient

Rehabilitation

Discharge

Fig. 7.2 Guidelines for fractured neck of femur in the first 24 h

one environment? It is ironic that so many policies and procedures are written with the aim of providing assurance and improving safety and yet the net effect is to degrade safety. They need to be drastically culled and simplified to produce a usable set of operating procedures analogous to those used in other high risk industries (Green et al. 2015).

Risk Control

Risk control strategies are used in healthcare in highly standardized and regulated environments such as pharmacy, blood products and radiotherapy where there are strict controls built into the delivery systems and restrictions on who can deliver therapies and what competencies they need. Risk control strategies could potentially be used much more widely particularly as a restraint on unnecessary or dangerous informal adaptation. Most importantly they could be used much more explicitly, with greater clarity and embraced as part of the patient safety armament. In this section we give examples of risk control strategies at both frontline and executive levels.

Control of Medication

Restrictions on the prescription and administration of drugs is a classic and widely used risk control strategy. For instance:

- There are clear guidelines about who can and cannot administer intrathecal chemotherapy (Franklin et al. 2014).
- Junior doctors are generally not permitted to prescribe certain drugs such as chemotherapy, oral methotrexate and other substances
- There are legal controls on the use of many drugs such as diamorphine and other opiates
- Nurses have to pass a test of competency to be permitted to administer intravenous medications

These restrictions are generally accepted but not thought of as a risk control strategy or as a patient safety initiative. We list them simply to make the point that risk control is already used and already accepted. The next example is rather different in being an example of the potential for risk control.

Potential for 'Go and No-Go' Controls in Surgery

Pre-flight checks require a conscious decision to proceed, referred to as a "go/no-go" decision. The civil aviation authorities set clear criteria governing the acceptable conditions for flying and it is expected that aircrew will recognise situations in which risk cannot be adequately managed. In such circumstances they are empowered to cancel the flight and indeed have a clear professional responsibility to do so. In contrast in healthcare the underlying assumption is to cope and carry on even in the face of considerable risk to patients. There are comparatively few areas in which 'no go' is explicitly understood and respected in healthcare.

National guidelines on equipment standards exist in anaesthesia. If faults are detected in core equipment it must be replaced, and if a suitable replacement is not available the case should not proceed without a specific, documented reason (Hartle et al. 2012). There are parallels between aviation and the operating theatre. An operation is a complex process that depends on the correct functioning of a number of different components, both human and technical. There are certain types of equipment failures in which it is assumed no anaesthetist would proceed (for example the airway gas analyser is unavailable), a situation in which some anaesthetists would proceed (an ultrasound is unavailable for a case requiring central venous cannulation), and a situation in which most anaesthetists might be expected to proceed (hospital uninterruptible power supply is unavailable, but all primary systems are functional). In practice however, although specific guidelines exist, there are very few clear 'no go' standards and the decision is left to the theatre team who are inevitably influenced by productivity pressures and other factors (Eichhorn 2012).

'No go' conditions could be defined in surgery to protect both patients and teams by imposing an inviolable limit which can only be bypassed in cases of emergency. 'No go' conditions are objective, absolute, minimum safety standards. They correspond to the thresholds above which activities of care must stop. The no go value correspond to a stage beyond which there is no capacity for safe care whatever the other strategies.

Placing Limits on Care

As we write this section in January 2015 a number of British hospitals have declared a 'major incident'. This does not necessarily relate to any specific incident but is a statement that they have reached crisis point and are unable to cope with the volume or type of patients they are receiving. This can happen in winter when demands are high, but also at other times, for example if there is a major road accident or a large number of older patients with pneumonia. This formal declaration allows the executive team to take a number of steps:

- One of the first measures is to start postponing routine activity, such as knee and hip operations or outpatient appointments.
- Cancelling leave and calling in more staff
- Making announcements to the public that the hospital is under pressure and not to attend the emergency department unless absolutely necessary

- In exceptional circumstances diverting ambulances so no emergency patients arrive. However, this is only used as a last resort as it increases demands on nearby sites.

This is a classic risk control strategy akin to grounding flights when an airport cannot cope with flight volume or in response to bad weather. Many hospitals take these measures in response to crisis but without necessarily having a clear cut prepared strategy in place. Risk control in its fullest sense though demands an explicit, preferably public approach to the problem to allow a considered strategic response rather than an ad hoc muddling through. Again, these critical strategies are not considered in the ambit of patient safety and are not studied, categorised, developed or taught.

Monitoring, Adaptation and Response

We have repeatedly emphasised that failures and departures from standards are not the exception but the day to day reality of healthcare. Safety is achieved partly by attempting to reduce and control such failures but also, in recognition of the impossibility of this task, by actively monitoring and managing problems that arise. The critical question is whether we leave this to ad hoc improvisation or try to build this capacity into the system (Vincent et al. 2013). Many proposed safety initiatives fall into this category but few have been implemented in a thoroughgoing and strategic manner. We provide some a small number of examples but there is huge scope for the development, formalisation, training and implementation of considered approaches to monitoring and adaptation.

Patients and Families as Problem Detectors

The active engagement and empowerment of patients and carers in an increasingly complex system poses huge challenges on many fronts. Patients and carers will have an increasingly important role in maintaining safety as home care expands, which will be discussed in the following chapter. At this point we simply want to highlight that almost all safety interventions that are aimed at patients fall into the category of monitoring, adaptation and response. In the hospital context patients and carers are in many cases being asked to compensate for problems of poor reliability and to form an additional defence against potential harm (Davis et al. 2011).

Many patient focused safety interventions are aimed at encouraging people to speak up if they notice problems with medicines, identification or other issues. More challengingly patients are asked to confront staff who have not washed their hands to support infection control (Pittet et al. 2011). Some of these interventions are entirely reasonable and in fact necessary; patients have a privileged and unique view of their own care and we need their insights into how safety is compromised. But we should be clear that patients are often being asked not only to check for problems that arise in complex care but to detect and compensate for problems that are not of their making.

Team Training in Monitoring, Adapting and Response

Teams, when working well, have the possibility of being safer than any one individual because a team can create additional defences against error by monitoring, double-checking and backing each other up: when one is struggling, another assists; when one makes an error, another picks it up (Vincent et al. 2010). Several authors have described how healthcare teams in emergency departments (Wears and Woods 2007) and operating theatres (Carthey et al. 2003) anticipate and thwart potential safety events. This can extend to more formal collaborative cross-checking, where one person, role, group or unit provides feedback about the viability or possible gaps in another's plans, decisions, or activities (Patterson et al. 2007). Allied to this is the development of a safety culture in which speaking openly about error is supported and indeed encouraged. Once one realises that errors and failures are inevitable, at least when the system is under pressure, the rationale for openness about error becomes clear. This kind of preparation is particularly critical in the more fluid and dynamic clinical environments where uncertainty is common and lapses frequent. For example, the WHO Surgical Safety Checklist is usually thought of as a means of checking processes such as the giving antibiotics in a timely fashion. However the checklist also prompts a brief period of reflection (the 'time out') in which members of the theatre team highlight potential problems and, by introducing each other, increases the chance of team members speaking up if problems are identified (Haynes et al. 2009; Kolbe et al. 2012).

Briefings and Debriefings, Handovers and Ward Rounds

Operational meetings, handovers, ward rounds and meetings with patients and carers are all sources of intelligence that allow the monitoring of safety For example, operational meetings held by senior managers can unblock beds and improve the flow of patients through a hospital, identify safety issues relating to infection outbreaks, and thwart the potential for unsafe discharge of patients. Briefings carried out by operating theatre teams provide an opportunity to identify and resolve equipment problems, staffing and theatre list order issues before a case starts. Debriefings carried out at the end of the theatre list support reflective learning on what went well and what could be done better tomorrow. Increasingly, briefings and debriefings are being introduced in other healthcare domains such mental health teams (Campbell et al. 2014).

Mitigation

The treatment and remediation of physical problems is obviously necessary when a patient has suffered some harm or complication. However psychological support is equally important both for patients and staff. Organisations vary hugely in the extent to which they are willing, prepared and able to provide support emotionally, practically and financially. Some hospitals have very well established systems for responding when patients have been harmed and highly developed mitigation strategies; others simply react and adapt.

Support Systems for Staff and Patients

The basic needs of injured patients have been understood for 20 years. We would all, in varying degrees, like an apology, an explanation, to know that steps had been taken to prevent recurrence and potentially financial and practical assistance (Vincent et al. 1994). We know that staff suffer a variety of consequences from being the 'second victim' as Albert Wu eloquently expressed it, not implying that the experiences of staff were necessarily comparable to those of injured patients (Wu 2000). We should also consider that a member of staff who has been seriously affected may well be performing poorly and be a risk to future patients; this again is rarely addressed. There are a few pioneering examples of programmes of support for both patients and staff (Box 7.2) but this is an area of safety management which needs substantial development (Iedema et al. 2011).

Box 7.2. Medically Induced Trauma Support Services (MITSS)
Linda Kenney, the founder of MITSS, experienced a grand mal seizure during an operation while cared for by an anaesthetist, Frederick van Pelt. Together they founded MITSS which provides support for both patients and staff. The Peer Support Programme uses colleagues as the primary support, following an approach that has been successfully used in the police, fire and emergency medical services. The programme aims to recruit credible, experienced clinical staff with personal understanding of the impact of error who are immediately available to provide confidential reflection and support. An education and training programme runs in parallel that aims to challenge the culture of denial of emotional response to serious errors and events. The hospital concerned made an active commitment to disclosure and apology and developed an Early Support Activation (ESA) programme for patients and families. The long-term strategy is to have a comprehensive emotional support for patients, families and care providers (van Pelt 2008).

The University of Michigan Health System pioneered a programme which included both support for patients and staff but also active intervention to provide compensation if appropriate and reduce the need for costly and potentially acrimonious litigation. The organisation performs active surveillance for medical errors, fully discloses errors to patients, and offers compensation when it is at fault. Evaluation of the programme found a decrease in new legal claims, number of lawsuits per month, time to claim resolution, and costs after implementation of the program of disclosure with offer of compensation. This approach did not increase legal claims and costs even in the notoriously litigious United States (Kachalia et al. 2010); in fact some decline in litigation was reported in Michigan generally through the latter part of the study period. Several New York hospitals have now implemented similar 'communication and resolution programmes'. To be successful they require the presence of a strong institutional champion, investment in developing and marketing the program to sceptical clinicians, and making it clear that the results of such transformative change will take time (Mello et al. 2014).

Regulatory and Political Determinants of Approaches to Safety

We have illustrated our five strategies within hospitals from the perspectives of both managers and frontline clinicians. To some degree they can determine the strategies they use to enhance safety. However they are also constrained by the wider regulatory and political environment. Regulators and politicians also have to decide on safety strategies for the wider system and their actions also determine the nature and feasibility of safety strategies within the organisations they influence. The two examples below show that the wider regulatory and political environment has a powerful influence not only on the form of healthcare that is delivered but on the safety strategies that can be adopted.

In France, the regulations governing radiotherapy, which are the province of the Nuclear Safety Agency (ASN), are much stricter than those governing the use of chemotherapy which is overseen by Haute Autorite de Sante (HAS). As a result, radiotherapists work to an ultra-safe model with many stipulations about the conditions of operation and an absolute requirement to minimise all errors and adverse events. ASN never hesitates to audit and suspend approval in cases of overdose or other serious problems. In contrast, oncologists have much greater freedom of action and are able to begin with a high dose (to bring maximum benefit) and reduce the dose as necessary depending on the patient's tolerance of unacceptable side effects. There are strict controls on the pharmaceutical production and on the preparation of chemotherapy, but comparatively few restraints on decisions about dose which are determined by the expert judgement of oncologists. These differences are in large part due to the different high-level requirements coming from the relevant authorities. Risk controls are

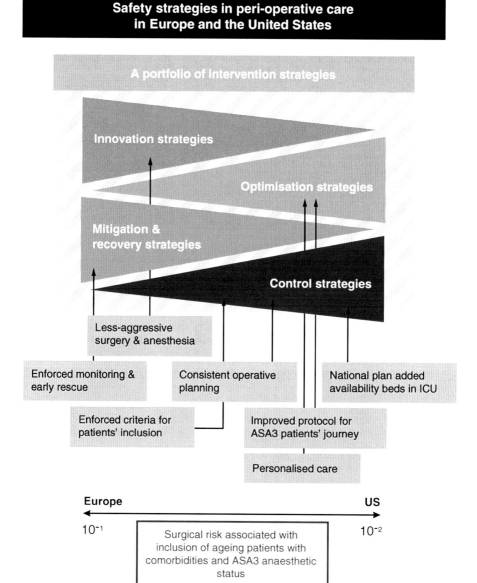

Fig. 7.3 Safety strategies in peri-operative care in Europe and the United States

imposed on radiotherapy, while much autonomy and adaptation is allowed for chemotherapy.

Different political contexts and levels of funding obviously influence the healthcare that can be delivered but also affect the safety strategies that can be employed. In this respect there are marked differences between approaches adopted in Europe and the United States in the surgical treatment of older patients with complex problems (Fig. 7.3). In Europe approximately 8.5 % of patients having major surgery are admitted to intensive care at some point in their hospital stay; mortality can be 4 % for all patients overall and as high as 20 % for older patients who are a poor anaesthetic risk. In contrast in the United States, 61 % of similar patients are admitted to intensive care; mortality is 2.1 % for all patients and 10–15 % for older patients with anaesthetic risk. These improvements in outcome in the United States are impressive but come at a considerable cost. In 2013, critical care services alone accounted for 4 % of all US health care expenditures, or nearly 1 % GDP (Neuman and Fleisher 2013). Europe has not made that choice which in turn means that different strategies must be employed which have a much stronger emphasis on the detection of problems and rapid response to mitigate the expected poorer outcomes (Fig. 7.3). In fact differences in mortality between high and low-volume hospitals are not associated with large differences in complication rates. Instead, these differences seem to be associated with the ability of a hospital to effectively rescue patients from complications. Strategies focusing on the timely recognition and management of complications once they occur may be essential to improving outcomes at low-volume hospitals (Ghaferi et al. 2009, 2011).

Safety in Context: The Many Hospital Environments

We have begun to set out the safety strategies that may be employed in hospitals, illustrating the broad strategies and the associated interventions. We recognise that much work is needed to explore this approach and map both actual and potential safety strategies and interventions. Another critical task is to consider how the strategies should be chosen and adapted to the many different environments within the hospital and in the light of the increasing complexity of care and the pressures on hospitals to provide safe care 24 h a day, 7 days a week.

We have previously argued that there are areas of the hospital which conform to our ultra-safe model, others which rely on a high reliability approach and a number in which care is highly adaptive, albeit still with a bedrock of core procedures. In some of these settings safety is best achieved by a mixture of automation, reliable equipment and adherence to core standards and procedures. In other environments these approaches remain important but need to be complemented by a greater reliance on risk control, adaptation and mitigation. Table 7.1 provides a general illustration of how we might employ different strategies to different contexts in the hospital as we develop the right blend of interventions and modes of operation. At the moment these ideas can only be proposed. However it would be possible, in fact necessary, to begin to identify and catalogue the strategies in day-to-day use using observational and ethnographic approaches and potentially quantify the reliance on them in different contexts.

Table 7.1 Choosing safety strategies in hospital care

Clinical context	Patient perspective	Safety strategies to adopt			
		Optimisation	Control	Adaptation	Innovation
Complex patients, limited protocols, frequent unexpected problems ICU, Emergencies, Oncology	Potential greater benefits as well as higher levels of risk	**Limited impact** Apply proven safety wins at the frontline. Simplify procedures and improve ergonomics	**Moderate impact** Delineation of competencies. Clear & effective rules for transfers to referral centres	**Big impact** Adapt staffing Celebrate Expertise Improve triage and access to the right person (expert) at the right time	**Lagged effect potential big impact** Adopt new standards, new drugs, new technology as soon as available
Scheduled care with usual disruptions (under staffing, organizational problems etc.) Standard scheduled surgery and medicine	Positive benefits with some risk	**Moderate impact** Apply proven safety wins Improve planning & organization. Simplify procedures and improve environment	**Moderate to big impact** Develop clear conditions for go/no go. Impose restrictions on patient inclusion/flow	**Big impact** Improve early detection and recovery of complications Improve teamwork safety culture and Patient participation	**Lagged effect potential moderate impact** Adopt proven new organisation, new techniques
Highly standardised care: radiotherapy, anaesthesiology	Little acceptance of error	**Big impact** Improve environment. Enforce protocols and audit compliance Monitor care and minimise side effects	**Big impact** Control patient selection. Careful pre-assessment Harmonize practices among carers. Enforce regulatory regime	**Moderate to limited impact** Priority given to prevention barriers	**Lagged effect limited impact** Priority given to stability and slow cycle of innovation
Medical support services: laboratory, blood, pharmacy	No acceptance of error	**Big impact** Enforce Protocols and legally binding rules. Increase audits	**Big impact** Enforce authorisations and regulatory regime	**Limited impact** Priority given to prevention barriers	**Limited immediate impact** Stability is an absolute priority for maintaining safety

Key Points

- In the past 15 years we can distinguish three phases of patient safety each associated with different types of action and intervention: the initial establishment of clinical risk management and the drive to reduce harmful incidents; a second phase in which industrial safety concepts and methods were applied to healthcare; a third phase of focal clinical interventions, team and cultural development. The earlier strategies have continued as the new ones emerged so that we now have 'a safety layer cake'.
- Many clinical services rely very heavily on ad hoc improvisation and adaptation to compensate for deficiencies of organisation and poor reliability of basic processes. The fact that a strategy is extensively used does not necessarily mean that it is desirable
- Safety may need to be approached differently in the varying clinical contexts within the hospital. All five strategic approaches will be needed in the hospital.
- Safety as best practice. Reduction of pressure ulcers, reduction of catheter-associated infections, improved hand hygiene, improved patient identification and adherence to core standards are critical in all environments.
- Improving the system includes standardising medication formularies and protocols, the introduction of information technology in all its forms, formalising roles and responsibilities in clinical teams, the use of care bundles and daily goals to organise ward care and all efforts to improve basic working conditions.
- Risk control includes: guidelines about who can and cannot administer intrathecal chemotherapy, legally controlled drugs with restrictions on their use and the implementation of go/no go conditions for surgical operations and other procedures
- Monitoring, adaptation and recovery includes: patients and families as problem detectors, teamwork and team training to adapt and recover, the use of briefings, debriefings and handover to anticipate and respond to problems.
- Mitigation. Organisations vary hugely in the extent to which they are willing, prepared and able to provide support emotionally, practically and financially. Some hospitals have very well established systems for responding when patients have been harmed and highly developed mitigation strategies.
- The wider regulatory, economic and political environment has a strong influence on the nature of the safety strategies that are feasible to employ within the healthcare system.

References

Association of Anaesthetists of Great Britain and Ireland (AAGBI), Hartle A, Anderson E, Bythell V, Gemmell L, Jones H, McIvor D, Pattinson A, Sim P, Walker I (2012) Checking anaesthetic equipment. Association of anaesthetists of Great Britain and Ireland. Anaesthesia 67(6):660–668

Bates DW (2000) Using information technology to reduce rates of medication errors in hospitals. BMJ 320(7237):788

Burnett S, Franklin BD, Moorthy K, Cooke MW, Vincent C (2011) How reliable are clinical systems in the UK NHS? A study of seven NHS organisations. BMJ Qual Saf. doi:10.1136/bmjqs-2011-000442

Campbell L, Reedy G, Tritschler C, Pathan J, Wilson C, Jabur Z, Luff A, Cross S (2014) Using simulation to promote safe and therapeutic services in mental health settings. BMJ Simul Technol Enhanc Learn 1(Suppl 1):A29

Carayon P (ed) (2006) Handbook of human factors and ergonomics in health care and patient safety. CRC Press, Boca Raton

Carthey J, de Leval MR, Wright DJ, Farewell VT, Reason JT (2003) Behavioural markers of surgical excellence. Saf Sci 41(5):409–425

Carthey J, Walker S, Deelchand V, Vincent C, Griffiths WH (2011) Breaking the rules: understanding non-compliance with policies and guidelines. BMJ 343:d5283

Davis RE, Sevdalis N, Vincent CA (2011) Patient involvement in patient safety: how willing are patients to participate? BMJ Qual Saf 20(1):108–114

Eichhorn JH (2012) The Anesthesia Patient Safety Foundation at 25: a pioneering success in safety, 25th anniversary provokes reflection, anticipation. Anaesth Analg 114(4):791–800

Flin R, Burns C, Mearns K, Yule S, Robertson EM (2006) Measuring safety climate in health care. Qual Saf Health Care 15(2):109–115

Franklin BD, Panesar SS, Vincent CA, Donaldson L (2014) Identifying systems failures in the pathway to a catastrophic event: an analysis of national incident report data relating to vinca alkaloids. BMJ Qual Saf 23(9):765–772

Ghaferi AA, Birkmeyer JD, Dimick JB (2009) Variation in hospital mortality associated with inpatient surgery. N Engl J Med 361(14):1368–1375

Ghaferi AA, Birkmeyer JD, Dimick JB (2011) Hospital volume and failure to rescue with high-risk surgery. Med Care 49(12):1076–1081

Green J, Evered R, Saffer J, Vincent CA (2015) Policies: less is more. Safe healthcare requires clear and simple operating procedures. BMJ (unpublished manuscript)

Haynes AB, Weiser TG, Berry WR, Lipsitz SR, Breizat AHS, Dellinger EP, Herbosa T, Joseph S, Kibatala PL, Lapitan MCM, Merry AF, Moorthy K, Reznick RK, Taylor B, Gawande AA (2009) A surgical safety checklist to reduce morbidity and mortality in a global population. N Engl J Med 360(5):491–499

Iedema R, Allen S, Britton K, Piper D, Baker A, Grbich C, Allan A, Jones L, Tuckett A, Williams A, Manias E, Gallagher TH (2011) Patients' and family members' views on how clinicians enact and how they should enact incident disclosure: the "100 patient stories" qualitative study. BMJ 343:d4423

Kachalia A, Kaufman SR, Boothman R, Anderson S, Welch K, Saint S, Rogers MA (2010) Liability claims and costs before and after implementation of a medical error disclosure program. Ann Intern Med 153(4):213–221

Kolbe M, Burtscher MJ, Wacker J, Grande B, Nohynkova R, Manser T, Spahn DR, Grote G (2012) Speaking up is related to better team performance in simulated anesthesia inductions: an observational study. Anaesth Analg 115(5):1099–1108

Mello MM, Senecal SK, Kuznetsov Y, Cohn JS (2014) Implementing hospital-based communication-and-resolution programs: lessons learned in New York City. Health Aff 33(1):30–38

Neuman MD, Fleisher LA (2013) Evaluating outcomes and costs in perioperative care. JAMA Surg 148(10):905–906

Patterson ES, Woods DD, Cook RI, Render ML (2007) Collaborative cross-checking to enhance resilience. Cognit Technol Work 9(3):155–162

Pittet D, Panesar SS, Wilson K, Longtin Y, Morris T, Allan V, Storr J, Cleary K, Donaldson L (2011) Involving the patient to ask about hospital hand hygiene: a National Patient Safety Agency feasibility study. J Hosp Infect 77(4):299–303

Pronovost PJ, Miller MR, Wachter RM (2006) Tracking progress in patient safety: an elusive target. JAMA 296(6):696–699

Reason J (1997) Managing the risk of organizational accidents. Ashgate, Aldershot

Reason JT, Carthey J, De Leval MR (2001) Diagnosing "vulnerable system syndrome": an essential prerequisite to effective risk management. Qual Health Care 10(Suppl 2):ii21–ii25

Sexton JB, Thomas EJ, Helmreich RL (2000) Error, stress, and teamwork in medicine and aviation: cross sectional surveys. BMJ 320(7237):745–749

Shekelle PG, Pronovost PJ, Wachter RM, Taylor SL, Dy SM, Foy R, Hempel S, McDonald KM, Ovretveit J, Rubenstein LV, Adams AS, Angood PB, Bates DW, Bickman L, Carayon P, Donaldson L, Duan N, Farley DO, Greenhalgh T, Haughom J, Lake ET, Lilford R, Lohr KN, Meyer GS, Miller MR, Neuhauser DV, Ryan G, Saint S, Shojania KG, Shortell SM, Stevens DP, Walshe K (2011) Advancing the science of patient safety. Ann Intern Med 154(10):693–696

Stanhope N, Crowley-Murphy M, Vincent C, O'Connor AM, Taylor-Adams SE (1999) An evaluation of adverse incident reporting. J Eval Clin Pract 5(1):5–12

Tsai TC, Joynt KE, Orav EJ, Gawande AA, Jha AK (2013) Variation in surgical-readmission rates and quality of hospital care. N Engl J Med 369(12):1134–1142

Van Pelt F (2008) Peer support: healthcare professionals supporting each other after adverse medical events. Qual Saf Health Care 17(4):249–252

Vincent C (ed) (1995) Clinical risk management. BMJ Publishing, London, pp 391–410

Vincent C, Phillips A, Young M (1994) Why do people sue doctors? A study of patients and relatives taking legal action. Lancet 343(8913):1609–1613

Vincent C, Taylor-Adams S, Stanhope N (1998) Framework for analysing risk and safety in clinical medicine. BMJ 316(7138):1154–1157

Vincent C, Taylor-Adams S, Chapman EJ, Hewett D, Prior S, Strange P, Tizzard A (2000) How to investigate and analyse clinical incidents: clinical risk unit and association of litigation and risk management protocol. BMJ 320(7237):777

Vincent C, Benn J, Hanna GB (2010) High reliability in health care. BMJ 340:c84

Vincent C, Burnett S, Carthey J (2013) The measurement and monitoring of safety. Health Foundation, London

Wachter RM (2010) Patient safety at ten: unmistakable progress, troubling gaps. Health Aff 29(1):165–173

Wears R, Vincent CA (2013) Relying on Resilience: Too Much of a Good Thing? In: Hollnagel E, Braithwaite J, Wears R (eds) Resilient Health Care. Ashgate, Farnham, pp 135–144

Wears RL, Woods DD (2007) Always adapting. Ann Emerg Med 50(5):517–519

Wu A (2000) Medical error: the second victim. BMJ 320:726–727

Safety Strategies for Care in the Home

8

Patient safety has evolved and developed in the context of hospital care. The understanding we have of the epidemiology of error and harm, the causes and contributory factors and the potential solutions are almost entirely hospital based. Safety in home care is likely to require different concepts, approaches and solutions. Safety in this context has however been barely addressed and yet care provided in the home will soon become the most important context for healthcare delivery.

The term 'home care' can encompass a variety of residential settings in which people are cared for by family, nurses and other professionals. In this chapter we use the term in a more restricted way to refer to the care of people in their own home, with varying degrees of informal and professional support. We focus on people with illnesses, usually chronic conditions, who are either living independently or being supported in their own homes by family or professional carers. Much healthcare is already delivered in the patient's home and this form of provision is growing rapidly. The benefits of home based care have been widely discussed, but the risks have not been fully articulated. In this chapter we first briefly summarise the background to the expansion of home care and then consider the nature and challenges for patient safety and the strategies that might help us manage risk in the home.

An Ageing Population and the Expansion of Home Care

More than 20 % of citizens in developed countries will be over 65 in 2020. These people, while enjoying better quality of life than previous generations, will suffer from a variety of long term conditions. As we discussed earlier, patients with cancer, heart disease, dementia, renal and respiratory disorders may now live for decades with their disease. The most common causes of disability however are due to sight and hearing disorders which affect very large numbers of people and are particularly pertinent to safety in the home. As well as an absolute increase in the numbers of older people, there will also be a considerable relative increase. The so-called 'support ratio' – the ratio of people of working age to those over 65 – will

© The Author(s) 2016
C. Vincent, R. Amalberti, *Safer Healthcare: Strategies for the Real World*,
DOI 10.1007/978-3-319-25559-0_8

decline substantially. Due to urbanization, migration and other factors frail older people will be more likely to live alone (United Nations Population Fund 2012).

Avoiding unnecessary hospitalization is a high priority for people living with chronic conditions. Once admitted to hospital, older adults are at an increased risk of poor outcomes such as readmission, increased length of stay, functional decline, iatrogenic complications and nursing home placement (Lang et al. 2008a; Hartgerink et al. 2014). The primary goals for care in the home are to avoid rehospitalisation and maintain a good quality of life.

A substantial growth of home care services appears to be inevitable. There has been a 50 % average increase of 'hospital at home' services in the past 10 years in Western countries and the rate is accelerating steeply. For instance in the United States 1.7 million people are currently employed as home care workers, with 7.2 million patients benefitting from these services. However the number of people receiving home care services is projected to rise to ten million by 2018 and to 34 million by 2030 (Gershon et al. 2012). This growing demand for the provision of nursing and rehabilitative care in the home as an alternative to hospital care contrasts with a scant literature on the safety, effectiveness and acceptability of hospital-at-home programmes, and evidence about their relative costs (Harris et al. 2005).

The Challenges of Delivering Healthcare in the Home

The familiar hospital model of healthcare delivery cannot easily be adapted to care delivered in the person's home. Patients are much more autonomous and coordination between professionals is much more difficult. Patients and carers play a much more active role and take on many responsibilities that are, in other settings, the prerogative of professionals. They may be responsible for care planning, for sharing relevant information with providers and for execution of care plans, including carrying out home monitoring and therapeutic regimens (Lorincz et al. 2011).

Patients and carers also have an important role in diagnosis and assessment, in that they must assess the seriousness of any change in condition and decide when, and how quickly, to escalate the response by bringing in other services. Their decisions may not concur with those made by the professionals involved (Barber 2002). Home care in all its forms needs to be negotiated to a much greater extent than in other settings in which professional values and organisation hold sway. In this context, patient preferences and values will often have a higher priority than medical guidelines and recommendations. Ultimately, it is the patients, their families and caregivers who decide what they will or will not do or accept (Stajduhar 2002).

To be at home is comforting for patients because of the familiarity of the environment and the trust in carers. The home looks very different to professionals who see multiple problems such as lack of knowledge, fall-inducing obstacles, unpackaged medications, misuse of proper disposal containers for syringe and needle and so on.

Professionals cannot determine the standard of safety independently of the recipients' perspectives, because such standards will have an impact not only on the patient but also on the lives of everyone involved.

While there is general agreement on the challenges of delivering care at home, there is huge variation in how different countries are responding to the challenge. In a recent seminar at the Institute of Healthcare Improvement (IHI 2014), United States representatives described a strategy of investing in the rapid development of information technology (such as tele-health and biosensors) as the ultimate solution for greater safety and efficiency of community and home care. In striking contrast, many other countries represented (particularly Japan, the Netherlands and Finland) were primarily aiming to improve solidarity among families and citizens, reduce disparities and refocus the role of doctors and nurses while maintaining affordable home care. Japan has trained "dementia supporters" who are expected to have the necessary knowledge and skills to support people with dementia and to create and promote a supportive culture for dementia. These different approaches make very different assumptions about how care is best managed but all will face major challenges in managing risk and maintaining safety.

The Hazards of Home Care: New Risks, New Challenges

In the last 20 years a series of studies have revealed the hazards of care in hospital. In consequence we tend to assume that patients will be safer at home; this is no doubt true for people who are relatively well, but may not be true for the frail and vulnerable. Care at home could, in some circumstances, generate even more adverse events than hospitals. The advancing age of the average patients at home and increasing numbers of comorbidities and medications are all associated with increased risk of experiencing a medication error or an adverse event (Lorincz et al. 2011). We cannot foresee all the potential hazards but studies are beginning to illuminate some of the dangers to patients and to carers.

Accidental Injury in the Home

Home is a more dangerous environment than most of us imagine. The leading causes of unintentional home injury deaths are falls, poisoning, fire and burns, airway obstruction, and drowning. Elderly residents are disproportionately affected, accounting for more than 2.3 million home injuries and 7000 unintentional home injury deaths annually in the United States (Gershon et al. 2012). People who are both old and ill are likely to be still more vulnerable to accidental injury. Risk factors include decline in physical or mental function, unsafe behaviours (such as smoking), living alone and health care management factors such as polypharmacy and lack of medication review (Doran et al. 2009).

Adverse Events in Home Care

An early study of home care in Canada found that 5.5 % of 279 home care clients suffered adverse events; injurious falls accounted for nearly half, followed by medication-related events, pressure ulcers and psychological harm (Johnson 2005). Two recent studies, one conducted in the USA (Madigan 2007) and the other in Canada (Sears et al. 2013), found that 13 % of home care patients experienced an adverse event. Larger estimates based on expert chart review of 1200 patients discharged in 2009–2010 in Canada showed a rate of 4.4 % adverse events (Blais et al. 2013). The most frequent were injuries from falls, wound infections, behavioural or mental health problems and adverse outcomes from medication errors. The number of comorbid conditions and the level of dependency greatly increased the risk of experiencing an adverse event. Patients can also be victims of abuse from family members, which might not always be readily apparent to care providers (MacDonald et al. 2011).

Adverse Drug Events

Adverse drug events have been the most studied safety issue in the home. Some studies have found that as many as 5 % of patients who were receiving nursing support at home had suffered from an adverse drug event of some kind during the previous week (Ellenbecker et al. 2004) and 25 % in the past 3 months (Sorensen et al. 2005). These problems are often due to poor communication between hospital staff, patients and their doctors in primary care (Ellenbecker et al. 2004). Few studies directly assess medication error caused by patients and family members, though models of human error should be equally applicable to patients and informal caregivers as to professionals (Barber 2002). In an Australian study, 35 % of readmissions were associated with incorrect drug administration at home. Those who had large stocks of medication at home were more exposed to adverse events (Sorensen et al. 2005). The majority of patients receiving home care services are taking more than five prescription drugs and over a third of patients are taking medications in ways that deviated from the prescribed medication regimen (Ellenbecker et al. 2004).

Risk to Family and Other Care Givers

Unpaid carers are particularly vulnerable to stress, long term burn out and ill health. Although health care aides play a role in giving assistance, the range of tasks falling to carers is considerable: assistance with eating, moving, washing, cleaning, connecting systems, improvising when systems fail, making decision on drug doses adjustments and responding to symptoms, often without any external advice or guidance.

Caring for a person with dementia is a full time occupation with no restriction on hours or oversight from the occupational health and safety regulations which protect professionals. Care at home is viewed positively as reducing the burden on the healthcare system; it might be more accurate to say that the burden is being transferred to the family and the patient themselves. The safety of professional care givers is also of concern, in that they are often sole workers who need to venture into dangerous areas to care for people who may themselves be dangerous. Increasing use of home care is bound to increase these risks, although these can be mitigated with proper support and appropriate technologies.

Problems of Transition and Coordination

The period following discharge from hospital is a particularly vulnerable time for patients. About half of adults experience a medical error after hospital discharge, and 19–23 % suffer an adverse event, most commonly an adverse drug event (Greenwald et al. 2007; Kripalani et al. 2007). Hospital discharge is poorly standardized and is characterized by discontinuity and fragmentation of care. At the time of first follow up with their primary doctors after hospitalization, up to 75 % of patients find that discharge summaries have not yet arrived which restricts their doctor's ability to provide adequate follow-up care (Schoen et al. 2012).

The above hazards illustrate some of the more obvious potential risks to patients and carers in the home environment. However the literature is not extensive and still primarily guided by a hospital based vision of adverse events. We are far from having a full picture of the combined benefits and risks of home care in relation to care provided in other settings.

Influences on Safety of Healthcare Delivered in the Home

Patient safety at home cannot be conceptualized or managed in the same way as patient safety in hospital because of the very different environment, roles, responsibilities, standards, supervision and regulatory context of home care. People are cared for in their homes and within the context of their family and the daily lives of all concerned. The quality and safety of care is influenced by the nature of formal service provision and the characteristics of the client receiving care, the physical environment and the availability of family and other carers (Hirdes et al. 2004; Lang et al. 2008b). We outline some of the main factors that will need to be assessed and understood when designing safe home care services.

Socio-economic Conditions Take on a Much Greater Importance

In an institutional setting, patients receive a certain standard of care regardless of their socioeconomic or cognitive status. In contrast, resources and environment of the home will vary hugely by socio-economic status. Wealthier people will be able to have a much higher standard of home care; they will have space for separate 'hospital' accommodation, paid support workers, leisure, better nutrition, less disruption of family life, and a higher probability that relatives can 'work' as carers. If a reasonable standard is to be achieved in poorer homes specific resources would have to be allocated to poorer families and to supporting the medical professionals in charge of those patients at higher social risk.

The elderly and disabled can be supported in their own environment 24 h a day by numerous 'smart' devices (Anker et al. 2011). Advances in telecommunication technologies have created new opportunities to provide tele medical care as an adjunct to medical management of patients. Feeling safer comes with a cost however, and that cost is often paid by the family. Contemporary homes are not typically designed or envisioned as places where complex or long-term health care is provided. The plethora of intrusive equipment, combined with the continual presence of carers, can make the person feel that their home is no longer a home.

The Home Environment as Risk Factor

The role of design in either degrading or promoting patient safety is increasingly understood. New hospitals may now be built with safety in mind, using good design to reduce equipment problems, assist infection control and reduce errors of all kinds (Reiling 2006). Once we move into the home, this hard won gain in understanding is largely lost. Stressful and potentially hazardous conditions, such as poor lighting, excessive clutter, presence of vermin, and aggressive family members, inadequate or unavailable sharps containers, and lack of readily accessible personal protective equipment, can directly or indirectly greatly increase the risk of adverse events in this population (Gershon et al. 2009, 2012).

In some homes performing clean or sterile procedures may be almost impossible. There is also the possibility that home care staff may transmit infections between homes, particularly when patients have been discharged after contacting MRSA or C-difficile. Hand washing provides some protection but cleaning equipment in the home environment is challenging (MacDonald et al. 2011).

The Household safety survey checklist (Table 8.1) includes the checking of fire and electrical risks, ergonomic (falls hazards), biological (unsanitary conditions), chemical, and other problems such as noise, temperature, poor security and violence. Additional items address various patient characteristics that influence safety. These include age, sex, health status, ability to walk without help, number of people in the household, daily medication, methods patients use to keep track of medications, presence of any medication in the home that patients no longer take, hearing aid use and the use of durable medical equipment and safety devices.

Increasing Responsibilities of Carers

Responsibility for safety at home largely falls on the shoulders of the patient, family members and informal carers. Caregivers are a particularly vulnerable group with

Table 8.1 Safety checklist for household hazardous conditions

Hazard categories	
Fall hazards	No non-slip mat in shower
	No grabs bars in shower or bath
	No nonslip rug on bathroom floor
	Loose or worn rugs or carpets
	Poor lighting
	Uneven or slippery floors
	Excessive clutter
	Awkwardly placed furniture
Fire and electrical hazards	No fire extinguisher
	No carbon monoxide alarm
	No smoke alarm
	Electrical cords damaged or overloaded
	Unsafe smoking materials
	Dangerous space heater
	Stove/cooker controls hard to reach
	Flammables near cooker top
Biological, hygiene and chemical hazards	Signs of cockroaches
	Signs of rats or mice in the home
	Excessive dust or animal hair
	Signs of lice, fleas or bed bugs
	Mould or fungus
	Rotten food or milk in the home
	Rubbish building up in the home
	Food not stored in a sanitary manner
	Cleaning products and other potential poisons are not in their original containers (original labels not in place)
Other miscellaneous hazards	No emergency contact list available (for family, doctor and others)
	Excessively load noise in the home (from either inside or outside)
	Doors lacking robust locks
	Threat of violence from aggressive dogs or other pets
	Threat of violence from neighbours
	Presence of weapons

Adapted from Gershon et al. (2012)

an increased risk for burnout, fatigue and depression. Some family members or friends work 24 h a day, 7 days a week, and a number of them try to continue their work outside the home. Family and other unpaid caregivers often make promises out of love and a sense of responsibility to keep the client at home, without being aware that this may be beyond their capacity (Stajduhar 2002).

The Training and Experience of Home Care Aides

Home care support workers play a significant role in maintaining safety at home. In the United States for example, with more than two million home healthcare employees and an anticipated employment growth of 48 % by 2018, the home healthcare workforce sector is the fastest growing in the U.S. healthcare system (Gershon et al. 2012). Home care aides help keep patients safe (Donelan et al. 2002) but they can also contribute to adverse events. Almost all are engaged in medication administration, but many lack knowledge of medicines. A Swedish study suggested that home care aides had a poor understanding of the hazards of the drugs they administer. Only 55 % knew the correct indications for common drugs and only 25 % knew the contraindications and symptoms of adverse drug reactions (Axelson and Elmstahl 2004).

Patients, family and even paid carers may all struggle to follow basic procedures which can be much more easily overseen and controlled in a hospital environment. We cannot rely on clear procedures and a strict regulatory environment for healthcare in the home. Both patients and health care aides are apt to rely on their capacity to muddle through and recover from errors. It is therefore important to acknowledge that recovery strategies (Johnson 2005) may be more important than prevention in the context of home care.

Fragmented Approach of Healthcare Professionals

Coordination and communication among providers and across organisations and sectors is a complex issue, especially vulnerable at the interfaces along the continuum of care (Romagnoli et al. 2013). As many as ten different professionals may be involved in the care of a patient in their home and each may be based in a different organisation and a different location. Coordination of care can be extremely problematic and there is considerable scope for the patient to receive conflicting or ambiguous recommendations which raise the risk of non-adherence and other safety issues.

In a recent UK survey, most patients expressed a preference for seeing a particular doctor, rising from 52 % among those aged 18–24 to over 80 % among those over 75. However, more than a quarter of patients reported being unable to see their preferred general practitioner consistently and recent evidence suggests that interpersonal continuity has declined in both inpatient and ambulatory care (Campbell et al. 2010; Sharma et al. 2009).

Safety Strategies and Interventions in the Home

Safety interventions in home care are challenging for professionals since they question usual assumptions and approaches. Priority is given to avoiding hospitalisation while increasing autonomy, and mental and social wellbeing. In this context, where there is often a trade-off between autonomy and safety, the best and safest care is a 'mastered compromise' in which a team of the patient, health and social care professionals and relatives each brings their own perspective and together arrive at a negotiated way forward. We believe however that, in addition to the thoughtful negotiation with patients and families, that it will also be valuable to consider broader strategic approaches to safety.

Optimization Strategies in Home Care: Best Practice and System Improvement

Optimization strategies are challenging to implement in the home especially with frail older people and people with mental health problems. The opportunities to directly implement evidence based medicine or to improve the delivery of care within the home are limited. Direct improvement of care can be difficult, time consuming and to reach only a proportion of the target group as the example in Box 8.1 shows.

> **Box 8.1. Difficult Challenge for Optimisation Strategies: Lessons from a Centralised Nurse-led Cholesterol-Lowering Programme**
> Lowering low-density lipoprotein (LDL) cholesterol in patients with diabetes mellitus (DM) and cardiovascular disease is critical to lowering morbidity and mortality. A team-based quality improvement programme attempted to improve compliance with evidence based medicine; registered nurses followed a detailed protocol to adjust cholesterol-lowering medications. General practitioners agreed to enrol 74 % of potential eligible patients. Thirty-six per cent of approved patients could not be reached via phone and 5.3 % declined enrolment. Of patients enrolled, 50 % did not complete the programme. Of those enrolled, median LDL decreased by 21 mg/dL and 52 % (33/64) achieved the LDL target.
> The resources required to identify, enrol and continually engage eligible patients raise many concerns regarding efficiency and highlight the challenges of implementing clinical guidelines in the home and community.
>
> Adapted from Kadehjian et al. (2014)

There are however important examples of successful initiatives which fall into the optimisation approach. Studies have examined the effectiveness of particular approaches to treatment at home, covering areas such as skin care and integrity, behaviour management, pain management and incontinence. The results of such research in nursing homes often show that "what works" involves simple, low-technology solutions that may increase staff time with patients (Stadnyk et al. 2011). In other words, the time staff spend listening to patients and carers, explaining, and coordinating may be one of the best ways of improving safety in the community and home care.

Discharge Planning and the Journey from Hospital to Home

Improving the patient journey from hospital to home and improving communication and coordination between professionals are critical in the support of patients returning home. Clear and timely hospital discharge information, including medication reconciliation, are key to this improvement. The advent of new professions such as care managers and practice facilitators in primary care is an important development in supporting patients at home with establishing personalized medical plans, coordination of professionals and the navigation of the healthcare system.

Patients at risk of poor outcomes after discharge may benefit from a comprehensive discharge planning protocol implemented by advanced practice nurses (Tibaldi et al. 2009; Shepperd et al. 2009); one in five hospitalizations is complicated by a post discharge adverse event. In one successful intervention, a nurse discharge advocate worked with patients during their hospital stay to arrange follow-up appointments, confirm medication reconciliation, and conduct patient education with an individualized instruction booklet that was sent to their primary care provider. A clinical pharmacist called patients 2–4 days after discharge to reinforce the discharge plan and review medications. Participants in the intervention group had a lower rate of subsequent hospital utilisation (Jack et al. 2009).

Training of Patients and Carers

Recently a member of one of our families had a cancer removed and was left with a substantial wound which needed regular dressing. The person was discharged home one day after a successful operation with the patient's partner, after minimal instruction, being responsible for the dressing of the wound, managing a drain and dealing with an incipient infection. This would, of course, have been unthinkable a few hours previously when the patient was in hospital. Fortunately the patient's partner proved adept at these rather difficult tasks. The early discharge was well intentioned and in the patient's best interest but the story illustrates how quickly professional standards are lost once the patient is discharged home.

In some settings, particularly in mental health, there is a much stronger emphasis on responsibility for the patient continuing beyond discharge and including preparation for return to home and life in the community. Physical healthcare is moving into the home and community but often without this mind-set of anticipation, preparation and continuing responsibility. If patients and carers are to take on essentially professional roles, albeit only with specific tasks, then surely they need to be trained to do so? In India, families have been co-opted as part of the workforce to help care for the patient but, in recognition of this role, they are prepared and trained (Box 8.2).

> **Box 8.2. Training Families to Deliver Care**
> At Narayana Health families are seen as having a crucial role in the recovery of patients following surgery. They operate a 'Care Companion Programme' to harness family members' potential and position them as an integral part of

the patient's recovery. A free structured training programme, tailored for those with low literacy levels, provides family members with simple medical skills such as monitoring vital signs, encouraging medicines adherence and supporting physical rehabilitation. The programme improves the quality and hours of care, leverages an untapped workforce, reduces costs and is universally transferable. Five thousand people a month are being trained on the programme. Given the desire to place patients and families at the centre of their care in the NHS, such training seems a practical way to help achieve it.

Adapted from Health Foundation (2014)

Risk Control Strategies in Home Care

Risk control strategies are difficult to impose in the home environment as much of the usual healthcare regulatory framework does not apply. We may however have to give some thought to a framework of standards and other controls as more healthcare is delivered in the home, particularly when patients live in isolated or poorer areas and need additional support to make home care a reasonable option.

There are almost no national standards regulating the physical environment in which home care services are provided, a stark contrast to requirements for healthcare institutions. Several household safety check lists have been developed to assess the compatibility of home with home hospitalization (Gershon et al. 2012). Imposing any restrictions may be difficult to achieve because any controls would require the full consent of the patient and family. Developing safety standards in the home presents a considerable challenge as hospital oriented approaches may have limited applicability in the home. Similar conflicts and difficulties may arise even in institutional home care settings (Box 8.3).

Box 8.3. Safety Standards in Home and Residential Care: Autonomy, Rights and Safety

In French hospitals there is a legal requirement that all medication should be given to patients by professionals. Patients cannot be entrusted with their own medication. Conditions for hospitalisation at home obviously differ from conditions in the hospital. In particular the autonomy of the patient and their carers is much greater. However French regulatory authorities, given the current law, have so far been reluctant to delegate taking medication to the patient. In practice patients at home are free to act as they choose regardless of the views of the regulatory authorities. Modifying this law will require an exception to be made for home hospitalisation, with the risk of increased ambiguity about the respective roles of patients and professionals.

Regulatory systems face considerable challenges in home care. For instance, French law considers that senior residents of retirement home no longer have a private home. Their bedroom in the residence is therefore considered as their home with all associated rights and privileges including adding personal furnishing, smoking, and even cooking. This was previously entirely positive as residents were entering retirement homes in their 80s while still able to live relatively independently. With an ageing population, and growing cost of retirement homes, people are more commonly entering retirement homes in their 90s and 70 % have severe cognitive impairment problems. The risk of fire when smoking, combined with limited medical access to the patient due to personal furniture, are now very high. The internal rules and regulations often forbid smoking and adding unsuitable furniture, but can be successfully challenged by patients and their relatives as a deprivation of rights. Changing these laws is not straightforward since this issue concerns a fundamental principle of freedom given by the French Constitution.

Monitoring, Adaptation and Response Strategies in Home Care

Monitoring, adaptation and response strategies are clearly to the fore as safety strategies for home care. The assumption that healthcare staff and organisations should wait for patients to present with an illness is giving way, at least for some chronic conditions, to a more proactive approach to monitoring, detection of problems and response aided by a variety of innovations in information technology.

In the hospital monitoring and detection of problems is largely the responsibility of staff. In the home however, patients and carers need to monitor, adapt and respond. This raises the question of how, as with staff, these abilities can be supported, encouraged and perhaps trained. This requires, as in other contexts, the development of a safety culture, and potentially other transferable routines such as safety briefings. For instance a colleague, who is a carer for a family member with serious mental health problems, has described how she and her husband have developed a routine of regular morning telephone calls in which they review the day, the support for the family member, any worrying symptoms, medication availability and other issues. This is, in essence, a safety briefing. Such systems could be developed in partnership with patients and carers and become an established safety strategy. As yet however, we do not know of any attempts to develop formal safety strategies for patients and carers at home, although there are many examples of individual patients developing their own ingenious and innovative approaches.

Detecting Deterioration

Carefully designed and implemented care management and tele health programs can improve safety and reduce health care spending (Baker et al. 2012). Many smart homes and remote monitoring solutions are emerging to support patients at home (Chan et al. 2009). The critical safety issue however is how to detect deterioration. In the context of hospitals David Bates and Eyal Zimmerman have argued that 'finding patients before they crash' is the next major opportunity to improve patient safety (Bates and Zimlichman 2014). In hospitals the primary tools to improve detection are the electronic health record, physiological sensors, decision analytics and mobile phones, with the assumption of a rapid clinical response once deterioration is identified. All these can potentially be employed in the home but implementation is far from straightforward.

The potential for home monitoring to improve the management of chronic conditions is considerable. Four of eleven programs that were part of the US Medicare Coordinated Care Demonstration reduced hospitalizations by 8–33 % among enrolees who had a high risk of near-term hospitalization (Brown et al. 2012). Home monitoring can come in the form of telephone support and visits, the promotion of self-care and the use of a variety of external or implantable devices. Multi-component interventions variously incorporate enhanced team communication, care planning, education and support for patients and carers, direct access to hospital care and the use of information technologies (Jaarsma et al. 2013). Tele medical monitoring service can combine with this support at home and reduce the number and duration of hospital admissions for worsening pathologies (Anker et al. 2011), though may not currently be suitable for patients with cognitive, visual or other sensory impairments (van den Berg et al. 2012). Implanted devices have been shown to be effective in reducing hospitalisation due to heart failure and reduce the need for active participation of the patient (Bui and Fonarow 2012).

> **Box 8.4. New Professional Roles Emerging**
> The care manager's central role is delivering and coordinating services for patients, including coordinating care across clinicians, settings, and conditions, and helping patients access and navigate the system. While these care coordination activities may benefit any patient, they can be particularly useful for those with chronic conditions and many care needs. Working closely with patients and their families, care managers' activities often include:
>
> - Assessing (and regularly reassessing) patients' care needs
> - Developing, reinforcing, and monitoring care plans
> - Providing education and encouraging self-management
> - Communicating information across clinicians and settings
> - Connecting patients to community resources and social services
>
> Adapted from Taylor et al. (2013)

It is becoming clear that successful home care requires not only monitoring but the development of a system of care which includes the selection of appropriate physiological indices, the timely interpretation of data by an experienced clinician, and a system capable of responding rapidly to provide appropriate treatment and to monitor the response to that treatment (Box 8.4). Few existing home monitoring approaches provide this full cycle of care and in addition these approaches will need to be tailored to individual patients according to disease severity, the patient's capacity for self-management, the availability of support and the home care environment (Bui and Fonarow 2012).

Mitigation

The benefits of providing healthcare in the home, for both minor and more serious conditions, are undoubted. As homecare becomes more complex however there will be a correspondingly greater risk of adverse events and therefore a need to anticipate and plan for a response to those events and mitigate their effects. In a hospital the rapid initiation of a remedial response to physical harm is part of routine clinical practice and we have previously discussed the need for psychological support for patients and staff. Mitigation strategies in the home will need to include consideration of both the psychological impact and preparation for an emergency response. In the event of a crisis the patient will need access to the right person at the right time; a capacity for rapid rehospitalisation whenever needed will be critical, especially at nights and week-ends.

The Responsibilities of Carers

The recognition that staff can be seriously affected by the role they have played in an error or harmful event has been a very important step forward, although programmes for supporting staff are still rare. In the home patients and carers are increasingly taking on professional roles and therefore they too may make serious and consequential errors. If a family member makes an error they have all the burden of responsibility that a professional bears combined with the terrible experience of harming someone close to them. Interviews with carers suggest that the responsibility for giving powerful medications can become burdensome both because of the time commitment and anxiety about making mistakes; many carers do not receive clear guidance about medication, leading to omissions, incorrect doses, anxiety and confusion which are often not recognised by health professionals (While et al. 2013). Relatives of people near the end of their lives face the additional worry about hastening death through improper use of medication (Payne et al. 2014). The blurring of boundaries between family carers and professionals is difficult for all concerned particularly towards the end of a person's life. As well as providing support and training to carers, we will also have to consider how to provide support in the event of a serious error, an issue which has currently not been addressed at all.

Mitigation Strategies in Home Haemodialysis

Home haemodialysis is hugely beneficial for patients in that dialysis at home preserves independence and autonomy and reduces dependence on the hospital. Patients and carers can become apprehensive about performing such a complex set of tasks and fearful about the potential for dialysis related emergencies (Pauly et al. 2015). Home dialysis is generally a very safe procedure but a number of deaths due to error have been recorded, such as a man who died from exsanguination after he connected a saline bag to a blood circuit (Allcock et al. 2012). In the early stages of home dialysis patients report frequent mistakes while they learn the procedures and develop their own personal safety strategies, such as ensuring that there are no interruptions and ensuring that help is on hand in the event of problems (Rajkomar et al. 2014).

Established haemodialysis units provide training and prepare patients and carers very carefully for home dialysis procedures. Instilling a culture of safety without unduly alarming the patient, ongoing vigilance from both patients and professionals and ongoing support are critical. In addition Pauly and colleagues (2015) suggest that it is necessary to develop safety strategies to mitigate the risk of adverse events, which include the anticipation and preparation for any adverse events that do occur. They set out a series of measures which includes the provision of an explanatory letter for a patient to take to an emergency department, ensuring the patient is fully briefed in emergency procedures, and a full set of procedures for staff to initiate to respond and learn from any events that do occur. The most important lesson from their account is the preparation that they provide for patients and carers includes an explicit and comprehensive set of safety strategies as part of the basic programme.

Reflections on Home Care Safety

By highlighting the risks of home care safety we do not intend in any way to suggest that care in the home is not desirable or possible. On the contrary it is essential for all of us who wish to live independently for as long as possible as we age. We can also see that innovations in remote medicine, tele monitoring and smart homes may well resolve some of the safety problems we have described. However care in the home does highlight some fundamental safety issues. Most importantly there is an apparent clash between autonomy and safety, although this is only a clash if you feel that older people must adhere to an ultra-safe model of safety. In reality safety is always only one of a number of objectives and we often knowingly take risks in the pursuit of other benefits, such as travelling, sport or exploration. More than that we accept the right of people to take personal risks even though the costs of failure often fall on the wider population when they are patched up again in hospital. Safety in the home needs to be assessed in the same way, not in terms of absolute safety but alongside other benefits. This is nicely captured in the term 'the dignity of risk' used in Australia by those providing services for frail older people. The model for safety in the home then is not ultra-safe; a frail older person at home has more in common with a deep sea trawler man than a pilot. Safety is managed by personal resilience, expertise and a high reliance on monitoring, adaptation and, most of all, recovery.

Key Points

- Safety in home care has barely been addressed and yet care given in a person's home will soon become the most important context for healthcare delivery.
- Many home care patients experience an adverse event. The most frequent adverse events are injuries from falls, wound infections, behavioural or mental health problems and adverse outcomes from medication errors.
- Patient safety at home cannot be conceptualized or managed in the same way as patient safety in hospital because of the very different environment, roles, responsibilities, standards, supervision and regulatory context of home care.
- Stressful and potentially hazardous conditions can directly or indirectly greatly increase the risk of adverse events at home
- Safety at home falls largely on the shoulders of patients, family members and relatives. Caregivers are a particularly vulnerable group with an increased risk for burnout, fatigue and depression.
- Limited available standards and the fragmented approach of healthcare professionals make home care more prone to errors
- Safety interventions must give priority to reduce hospitalisations, increase wellbeing, increase communication among carers and with patients, and improve recovery strategies.
- There are opportunities to implement evidence based care in the home but it is considerably more difficult than in other settings. Much can be done to improve support systems, detection of problems and recovery.
- Highlighting the risks of home care does not imply that care in the home is not desirable or possible. We should not aim for absolute safety in home care but assess risks in the context of the benefits of living as independently as possible at home. Safety is managed by personal resilience, expertise and a high reliance on monitoring, adaptation and, most of all, recovery.

References

Allcock K, Jagannathan B, Hood C, Marshall M (2012) Exsanguination of a home haemodialysis patient as a result of misconnected blood-lines during the wash back procedure: a case report. BMC Nephrol 13(1):28

Anker SD, Koehler F, Abraham WT (2011) Telemedicine and remote management of patients with heart failure. Lancet 378(9792):731–739

Axelsson J, Elmståhl S (2004) Home care aides in the administration of medication. International J Qual Health Care 16(3):237–243

Baker A, Leak P, Ritchie LD, Lee AJ, Fielding S (2012) Anticipatory care planning and integration: a primary care pilot study aimed at reducing unplanned hospitalisation. Br J Gen Prac 62(595):e113–e120

Barber N (2002) Should we consider non-compliance a medical error? Qual Saf Health Care 11(1):81–84

Bates DW, Zimlichman E (2014) Finding patients before they crash: the next major opportunity to improve patient safety. BMJ Qual Saf. doi:10.1136/bmjqs-2014-003499

Blais R, Sears NA, Doran D, Baker GR, Macdonald M, Mitchell L, Thalès S (2013) Assessing adverse events among home care clients in three Canadian provinces using chart review. BMJ Qual Saf. doi:10.1136/bmjqs-2013-002039

Brown RS, Peikes D, Peterson G, Schore J, Razafindrakoto CM (2012) Six features of medicare coordinated care demonstration programs that cut hospital admissions of high-risk patients. Health Aff 31(6):1156–1166

Bui AL, Fonarow GC (2012) Home monitoring for heart failure management. J Am Coll Cardiol 59(2):97–104. doi:10.1016/j.jacc.2011.09.044

Campbell SM, Kontopantelis E, Reeves D, Valderas JM, Gaehl E, Small N, Roland MO (2010) Changes in patient experiences of primary care during health service reforms in England between 2003 and 2007. Ann Fam Med 8(6):499–506

Chan M, Campo E, Estève D, Fourniols JY (2009) Smart homes—current features and future perspectives. Maturitas 64(2):90–97

Donelan K, Hill CA, Hoffman C, Scoles K, Feldman PH, Levine C, Gould D (2002) Challenged to care: informal caregivers in a changing health system. Health Aff 21(4):222–231

Doran DM, Hirdes J, Blais R, Ross Baker G, Pickard J, Jantzi M (2009) The nature of safety problems among Canadian homecare clients: evidence from the RAI-HC reporting system. J Nurs Manag 17(2):165–174

Ellenbecker CH, Frazier SC, Verney S (2004) Nurses' observations and experiences of problems and adverse effects of medication management in home care. Geriatr Nurs 25(3):164–170

Gershon RR, Pearson JM, Sherman MF, Samar SM, Canton AN, Stone PW (2009) The prevalence and risk factors for percutaneous injuries in registered nurses in the home health care sector. Am J Infect Control 37(7):525–533

Gershon RR, Dailey M, Magda LA, Riley HE, Conolly J, Silver A (2012) Safety in the home healthcare sector: development of a new household safety checklist. J Patient Saf 8(2):51–59

Greenwald JL, Denham CR, Jack BW (2007) The hospital discharge: a review of a high risk care transition with highlights of a reengineered discharge process. J Patient Saf 3(2):97–106

Harris R, Ashton T, Broad J, Connolly G, Richmond D (2005) The effectiveness, acceptability and cost of a hospital-at-home service compared with acute hospital care: a randomized controlled trial. J Health Serv Res Policy 10(3):158–166

Hartgerink JM, Cramm JM, Bakker TJ, Eijsden RA, Mackenbach JP, Nieboer AP (2014) The importance of relational coordination for integrated care delivery to older patients in the hospital. J Nurs Manag 22(2):248–256

Hirdes JP, Fries BE, Morris JN, Ikegami N, Zimmerman D, Dalby DM, Aliaga P, Hammer S, Jones R (2004) Home care quality indicators (HCQIs) based on the MDS-HC. Gerontologist 44(5):665–679

IHI (2014) Ageing IHI-ISQUA seminar, Boston, USA, November 2014

Jaarsma T, Brons M, Kraai I, Luttik ML, Stromberg A (2013) Components of heart failure management inhomecare;aliteraturereview.EurJCardiovascNurs12(3):230–241.doi:10.1177/1474515112449539

Jack BW, Chetty VK, Anthony D, Greenwald JL, Sanchez GM, Johnson AE, Forsythe SR, O'Donnell JK, Paasche-Orlow MK, Manasseh C, Martin S, Culpepper L (2009) A reengineered hospital discharge program to decrease rehospitalisation: a randomized trial. Ann Intern Med 150(3):178–187

Johnson KG (2005) Adverse events among Winnipeg home care clients. Healthc Q (Toronto, Ont) 9:127–134

Kadehjian EK, Schneider L, Greenberg JO, Dudley J, Kachalia A (2014) Challenges to implementing expanded team models: lessons from a centralised nurse-led cholesterol-lowering programme. BMJ Qual Saf 23(4):338–345

Kripalani S, LeFevre F, Phillips CO, Williams MV, Basaviah P, Baker DW (2007) Deficits in communication and information transfer between hospital-based and primary care physicians: implications for patient safety and continuity of care. JAMA 297(8):831–841

Lang A, Edwards N, Fleiszer A (2008a) Safety in home care: a broadened perspective of patient safety. International J Qual Health Care 20(2):130–135

Lang A, Macdonald M, Storch J, Elliott K, Stevenson L, Lacroix H, Donaldson S, Corsini-Munt S, Francis F, Curry C (2008b) Home care safety perspectives from clients, family members, caregivers and paid providers. Healthc Q (Toronto, Ont) 12:97–101

Lorincz CY, Drazen E, Sokol PE, Neerukonda KV, Metzger J, Toepp MC, Maul L, Classen DC, Wynia MK (2011) Research in ambulatory patient safety 2000–2010: a 10-year review. American Medical Association, Chicago

Macdonald M, Lang A, MacDonald JA (2011) Mapping a research agenda for home care safety: perspectives from researchers, providers, and decision makers. Can J Aging 30(02):233–245

Madigan EA (2007) A description of adverse events in home healthcare. Home Healthc Nurse 25(3):191–197

Pauly RP, Eastwood DO, Marshall MR (2015) Patient safety in home haemodialysis: quality assurance and serious adverse events in the home setting. Hemodial Int 19:S59–S70. doi:10.1111/hdi.12248

Payne S, Turner M, Seamark D, Thomas C, Brearley S, Wang X, Blake S, Milligan C (2014) Managing end of life medications at home-accounts of bereaved family carers: a qualitative interview study. BMJ Support Palliat Care 5(2):181–188. doi:10.1136/bmjspcare-2014-000658

Rajkomar A, Farrington K, Mayer A, Walker D, Blandford A (2014) Patients' and carers' experiences of interacting with home haemodialysis technology: implications for quality and safety. BMC Nephrol 15(1):195

Reiling J (2006) Safe design of healthcare facilities. Qual Saf Health Care 15(suppl 1):i34–i40

Romagnoli KM, Handler SM, Ligons FM, Hochheiser H (2013) Home-care nurses' perceptions of unmet information needs and communication difficulties of older patients in the immediate post-hospital discharge period. BMJ Qual Saf 22:324–332. doi:10.1136/bmjqs-2012-001207

Schoen C, Osborn R, Squires D, Doty M, Rasmussen P, Pierson R, Applebaum S (2012) A survey of primary care doctors in ten countries shows progress in use of health information technology, less in other areas. Health Aff 31(12):2805–2816

Sears N, Baker GR, Barnsley J, Shortt S (2013) The incidence of adverse events among home care patients. International J Qual Health Care 25(1):16–28

Sharma G, Fletcher KE, Zhang D, Kuo YF, Freeman JL, Goodwin JS (2009) Continuity of outpatient and inpatient care by primary care physicians for hospitalized older adults. JAMA 301(16):1671–1680

Shepperd S, Doll H, Angus RM, Clarke MJ, Iliffe S, Kalra L, Ricauda NA, Tibaldi V, Wilson AD (2009) Avoiding hospital admission through provision of hospital care at home: a systematic review and meta-analysis of individual patient data. Can Med Assoc J 180(2):175–182

Sorensen L, Stokes JA, Purdie DM, Woodward M, Roberts MS (2005) Medication management at home: medication-related risk factors associated with poor health outcomes. Age Ageing 34(6):626–632

Stadnyk RL, Lauckner H, Clarke B (2011) Improving quality of care in nursing homes: what works? Can Med Assoc J 183(11):1238–1239

Stajduhar KI (2002) Examining the perspectives of family members involved in the delivery of palliative care at home. J Palliat Care 19(1):27–35

Taylor EF, Machta RM, Meyers DS, Genevro J, Peikes DN (2013) Enhancing the primary care team to provide redesigned care: the roles of practice facilitators and care managers. Ann Fam Med 11(1):80–83

Tibaldi V, Isaia G, Scarafiotti C, Gariglio F, Zanocchi M, Bo M, Bergerone S, Ricauda NA (2009) Hospital at home for elderly patients with acute decompensation of chronic heart failure: a prospective randomized controlled trial. Arch Intern Med 169(17):1569–1575

United Nations Population Fund (2012) Ageing in the twenty-first century. New York. http://www.unfpa.org/sites/default/files/pub-pdf/Ageing%20report.pdf

van den Berg N, Schumann M, Kraft K, Hoffmann W (2012) Telemedicine and telecare for older patients—a systematic review. Maturitas 73(2):94–114. doi:http://dx.doi.org/10.1016/j.maturitas.2012.06.010

What can the UK learn from health innovation in India (2014). Health Foundation, London

While C, Duane F, Beanland C, Koch S (2013) Medication management: the perspectives of people with dementia and family carers. Dementia 12(6):734–750. doi:10.1177/1471301212444056

Safety Strategies in Primary Care

9

Patient safety is a young discipline that emerged from medico-legal concerns associated with the risk of occurrence of specific and easily identifiable adverse events that were mostly associated with hospital care. In primary care however patients are managed over long periods of time and the safety issues that arise are likely to be of a very different character. We have earlier suggested that we should recast patient safety as the management of risk over time; this perspective may be better adapted to the longer time scales of primary care.

With the exception of exceptional criminal behaviour, such as the example of Harold Shipman (Baker and Hurwitz 2009), primary care has not been considered as an important source of specific adverse events. The priorities in primary care have been to improve access and overall quality of care, rather than to examine system vulnerabilities and safety issues. However once we begin to examine safety over time, rather than in terms of specific incidents, safety issues may become more visible. In this chapter we briefly outline current knowledge of patient safety in primary care and then consider whether the five strategic approaches can be applied in this context.

Challenges for Primary Care

Primary care in every country faces huge challenges. People are living longer, often with one or more chronic conditions, and need a greater degree of support in the community while still expecting to have a good quality of life at home. Primary care practitioners are dealing with more patients with complex conditions and comorbidities making it impossible to provide the best and safest care to every patient. Primary care clinics have to coordinate both a very wide range of professions and respond to patient values and preferences. The increasing need to personalise medicine and engage the patient in decisions about their care, while according with the values of primary care, demands more time than is realistically available (Snowdon et al. 2014)

© The Author(s) 2016
C. Vincent, R. Amalberti, *Safer Healthcare: Strategies for the Real World*,
DOI 10.1007/978-3-319-25559-0_9

Primary care physicians express frustration that the knowledge and skills they are expected to master exceed the limits of human capability (Bodenheimer 2006). The introduction of genomics and personalized medicine will only increase the complexity and demands placed on primary care services and the knowledge and technologies that staff need to understand and employ. The number of general practitioners working alone is falling rapidly. Primary care doctors are working in larger clinics and federations to provide a more consistent and coordinated approach to care. Nurses and other professions are taking on increased responsibilities and a wider clinical remit. However these changes, while important, will not be sufficient to address current and future challenges. Safety in primary care needs to be reconsidered in the light of the above, increasing the priority of national primary care patient safety strategies and developing interventions appropriate to the context.

The Nature of Risk in Primary Care

Doctors in primary care work together to present and solve problems in short consultations, typically 7–16 min across Europe. Patients often (but not invariably) present with early manifestations of illness, often against a backgrounds of pre-existing psychosocial problems and physical co-morbidities. Diagnosis in such circumstances is necessarily provisional and general practitioners face an enormously difficult task in identifying the few cases of serious illness amongst the very large number of minor problems. To be 'safe' in this context, in the sense of being certain that a patient does not have a serious illness, is not feasible. To investigate every problem to achieve diagnostic certainty would not be good practice; the anxiety generated, the risks of investigation and tests and the inconvenience to patients would be counter-productive. In addition any healthcare system would be bankrupt within months. Given this equation, time is often used as a diagnostic and therapeutic tool, but always with considerable latitude.

Patients in primary care are much freer than in any hospital system. They may decide not to comply with their nurse or doctor's recommendations because they conflict with personal aims or lifestyle; this is typically the case of 30–50 % of patients (Barber 2002). Patients in primary care, because of their greater autonomy, may increase the risks of adverse events in some circumstances which poses many difficult ethical and medico-legal issues (Buetow et al. 2009).

Until recently many general practitioners worked alone or in small groups. This model of practice, often combined with a very high workload, made it difficult to see risk at a system level or consider broad risk management strategies. General practitioners and other primary care staff may have high personal standards of care without being aware of the frequency or impact of any errors or the vulnerabilities and risks to patients in the wider system of care (Jacobson et al. 2003). The flexibility, diversity and personal approach for every patient that primary care clinicians rightly regard as a strength make it very challenging to define error and adverse events in a reasonable and consistent manner.

Error and Harm in Primary Care

Studies in hospitals have shown that different methods of gathering data reveal different types of error and harm and that a combination of methods is needed to map the landscape of safety (Hogan et al. 2008). The same is true in primary care (Sandars and Esmail 2003). One study used five contrasting methods to identify adverse events: physician reported adverse events, pharmacist reported adverse events, patients' experiences of adverse events, assessment of a random sample of medical records, and assessment of all deceased patients. There was almost no overlap of adverse events detected between these methods. The patient survey accounted for the highest number of events and the pharmacist reports for the lowest number (Wetzels et al. 2008). These difficulties in measurement are partly due to the lack of developed systems of monitoring safety in this context but also to the difficulties of definition of both error and adverse events.

The top five medical errors reported by family physicians are: errors in prescribing medications; errors in getting the right laboratory test done for the right patient at the right time; filing system errors; errors in dispensing medications; and errors in responding to abnormal laboratory test results. Poor communication and coordination between professionals and different elements of the health and social care system are the primary cause of many of the problems identified (Dovey et al. 2003). The lack of timely and accurate information after patients are discharged from hospital and delays in obtaining test results are both major risks (Kripalani et al. 2007; Callen et al. 2012). A more recent study of adverse events in primary care (ESPRIT) used a prospective method gathering data over seven consecutive days (Kret and Michel 2013). General practitioners reported 475 errors over a total of 13,438 visits (just under 3 %) but 95 % of those reported errors were minor and any consequences they had were immediately recognized. These studies identify important problems, but they are restricted to those immediately visible to the primary care doctor, which in effect means those occurring within the clinic or involving communication with other services

Studies which monitor errors within a specific time period, while valuable, will clearly not detect problems that are only revealed in the longer term, such as wrong or delayed diagnosis which are far more prominent in analyses of claims and complaints. The most common allegation in medical negligence claims in primary care by far (up to 40 % of the total claims) is missed or delayed diagnosis especially for cancer and cardiac disease (Gandhi et al. 2006; Singh et al. 2013). This reflects again how difficult it is for general practitioners individually to monitor and detect rare but serious problems that are not immediately apparent in the daily routine and also the need to consider safety issues over much longer time periods in this context

Diagnostic Errors

Diagnostic errors have not yet received the attention they deserve, considering their probable importance in leading to harm or sub-standard treatment for patients; the emphasis on systems has led us away from examining core clinical skills such as diagnosis and decision making (Wachter 2010) but these are now becoming a major focus. Cancer outcomes in the United Kingdom, while improving, are not as good as in many European countries and this may be partly explained by delayed or incorrect initial diagnoses (Lyratzopoulos et al. 2014). Diagnostic errors are difficult to study, being hard to define, hard to specify as occurring at a particular point in time and not directly observable. The term 'diagnostic error' may indicate either a relatively discrete event, such as missing a fracture on an X-ray, or a narrative which unfolds over months or even years, such as a delayed diagnosis of lung cancer because of failures in the coordination of outpatient care (Vincent 2010). These examples show that the term error can be an oversimplification of a long story of undiagnosed illness.

Studies of multiple consultations in the presentation of cancer show that the nature of the disease, both its presentation and rarity, is a powerful predictor of speed of diagnosis. Most patients with cancer present to primary care with symptoms that have low or very low positive predictive values. Even "red flag" symptoms (such as rectal bleeding, dysphagia, haemoptysis, and haematuria) are not strongly associated with the presence of cancer. Despite these challenges about 80 % of patients with cancer are referred to a hospital specialist after one (50 %) or two (30 %) consultations. But a substantial minority (20 %) of patients with cancer visit a primary care doctor with relevant symptoms three or more times before referral. This number is often considered by policy makers and cancer charities to reflect an avoidable delay. These patients however are often those with cancers which are particularly difficult to diagnose because of their non-specific symptom pattern (Lyratzopoulos et al. 2014).

We still have many challenges to address even in providing a complete account of the various errors, adverse events and wider safety issues in primary care. Problems of definition, methodology and method abound. There is however evidence from a number of quarters of risks to patients from vulnerabilities in both individuals and systems, though this realisation must be tempered by the fact that primary care practitioners cannot (and emphatically should not) try to minimise all possible risk; such an approach would lead to massive over-investigation and treatment and would be completely unaffordable. Managing risk in this context is a challenging affair and we would suggest, currently largely conceptualised as the responsibility of individual doctors. Doctors, nurses and other primary care professionals obviously play a critical role in the management of risk in the negotiation, shared decision making and treatment of individual patients. However we need, as in hospitals, to look beyond the individual perspective and try to imagine what managing risk and safety across a population of patients might look like. Can we apply the framework of five strategies and the associated interventions to this context to provide a conceptual and practical approach to risk management in primary care?

Safety as Best Practice

Adherence to best practice and evidence based medicine is as important in primary care as in other contexts. In England considerable emphasis has been placed on external incentives for improving primary care most notably in the use of

Payment-for-Performance system (P4P). The United Kingdom pioneered the idea in 2004 with the Quality and Outcomes Framework, and the United States, France and other countries have developed similar schemes. The Idea of P4P is simple enough: pay for compliance to evidence based medicine. Pay for performance can certainly drive change in specific practices but its overall impact on quality of care and professional values is still debated (Lee et al. 2012; Hussey et al. 2011; Ryan et al. 2015). Irrespective of this, the more difficult issue from the safety perspective is that P4P does not address the three top three adverse events as cited in the literature: delayed and missed diagnosis; medication safety; and poor strategies of care and inadequate surveillance (Brami and Amalberti 2010; Lorincz et al. 2011).

Many interventions use quality improvement approaches to improve adherence to guidelines to improve outcomes for patients (Marshall et al. 2013). For example, depression in primary care settings is often not well managed or treated with resultant poor outcomes. As depression is one of the major causes of disability worldwide this is a critical issue. In a remarkable early study of quality improvement approaches, managed primary care practices in the United States were randomized to usual care or a quality improvement programme. The intervention involved institutional commitment to quality improvement, identification of a pool of potentially depressed patients, training local experts and nurse specialists to provide clinician and patient education, and either nurses for medication follow-up or access to trained psychotherapists. Mental health outcomes and retention of employment of depressed patients improved over a year, while medical visits did not increase overall. A modest investment in quality improvement produced substantial gains in some areas, including a marked increased detection of patients with depression (Wells et al. 2000).

Studies of known diagnostic errors in primary find that most concern common conditions such as pneumonia, cancer, congestive heart failure, acute renal failure, and urinary tract infections; this is of course partly because these conditions are common in any case. Problems identified may lie in the clinical encounter but are also related to referrals, patient-related factors, follow-up and tracking of diagnostic information, and performance and interpretation of diagnostic tests. While some of these problems may be addressed by improving the skills of individual practitioners this is unlikely to have a major impact. To begin with many diagnostic errors may be due to fundamental features of human cognition which are hard to change. People make frequent and effective use of heuristics in day to day thinking which are generally extremely useful but which can also mislead in situations where more analytical thinking is required (Kahneman 2011). Large numbers of heuristics and biases have been identified (Croskerry 2013) and it is not yet clear whether it is possible, still less cost effective, to train people to improve diagnostic accuracy. In terms of the management of risk we may be better to invest in improving the more tractable aspects of the system (such as communication of test results) and, probably even more important, investing more time and effort in following up patients who attend for an initial presentation with potentially serious symptoms. This would require the development of failsafe systems for overlooked abnormal tests and recall of patients who did not attend planned investigations or follow up appointments (Lyratzopoulos et al. 2014). In our terms we would move from a strategy of best practice and system improvement towards one of monitoring, adaptation and recovery. We would accept that some diagnostic delays and errors are inevitable and shift the balance of resources towards rapid detection.

Improving the System

Delivering the care suggested by guidelines is obviously a desirable objective, but the goal will remain difficult to achieve. There are many reasons for this but two are particularly important. First, guidelines are only a partial guide to treatment even for a relatively healthy person with a single condition. When caring for a frail older person with multiple problems doctors need to make many adjustments to achieve the best care for that individual (Persell et al. 2010). A second major problem is the extraordinary pace of medical innovation and the accompanying exponential growth in scientific knowledge in modern medicine; the half-life of knowledge is only 6 years in most specialities (Shojania et al. 2007; Alderson et al. 2014). Once the new knowledge is available, it takes time to consolidate and put it in the form of guidelines and recommendations. It takes as long again to establish a system of new recommendations to work, updated at a proper pace, and with relevant information for ageing comorbid patients.

The introduction of information technology and Electronic Health records represent the best chance of responding to the rapid evolution of medical knowledge and practice. New technologies are expected to assist and support medical decision making and prescribing, provide prompts for ordering and checking test results, enhance cooperation and allow patients to access their medical record (De Lusignan et al. 2014); they may also facilitate new approaches to measuring clinical performance and detecting poor care (Weiner et al. 2012). But the effective use of such technology will depend on consistent deployment at a national level and on associated training in how to use these systems effectively without being burdened by an overload of information, recommendations and alerts (Shoen et al. 2012; Jones et al. 2014). The full benefits of such systems have not yet been realised but we are already beginning to see that their introduction has a number of unanticipated consequences, some of which are highly undesirable. For instance clinicians in the past would go to speak to a radiologist to discuss an ambiguous CT scan, whereas now they will make their own solitary decision from a screen. Young doctors will review their patients on the electronic health record rather than actually go and see them (Wachter 2015).

Optimisation strategies (best practice and improving the system) are perfectly feasible when conducted at scale but much more difficult to implement in a small primary care clinic or practice. The improvements needed in the primary care system and the levers of change are national or at least regional issues. Programmes with clear and specific improvement targets can have an impact, as the example of treatment of depression shows, but considerable resources are needed to have an impact at scale. Primary care clinics can make some use of optimisation strategies but, at a local level, may need to make more use of the other forms of safety strategy which place a stronger emphasis on the active management of risk.

Risk Control Strategies

When systems are under pressure risk control strategies need to be considered to maintain safety and potentially also to constrain costs. An important example of risk control is the deliberate restriction of clinical practice in circumstances posing high risk to patients. This method of controlling risk is most prominent in the prescription and administration of high risk medication. For instance, in primary care certain drugs cannot be prescribed by general practitioners or supplied by community pharmacists.

The argument for risk control is essentially that it is better to explicitly manage demand and conditions of work in order to maintain standards and preserve safety; the alternative is a system which delivers some high quality care but which potentially runs out of control. The most obvious potential control is to cap the maximum number of patients who are under the care of a single primary care team, which varies according to patient characteristics and how care is provided. Provided some realistic assessment of patient need and consultation length can be made, then capping the number of patients per team is a possible option but solutions of this kind can only be achieved at a national level. Demand can be managed locally by greater involvement of nurses and paramedical staff in care delivery or by transfer to hospitals or other facilities but this may not be feasible in isolated or poorer areas. There are however many examples of risk control in primary care and the potential for a much more thorough consideration of this particular strategy.

Control by Assessment of Competency

Almost all countries with developed healthcare systems have procedures for licensing doctors, identifying and potentially retraining those who fall below the required standard. Whatever the merits of the a systems approach to safety, there is no doubt that a proportion of problems are linked to the standard of care provided by individuals; in Australia for instance 3 % of the workforce accounted for half the complaints made with some individuals the subject of repeated complaints. Many countries require general practitioners to engage in a CPD or CPE programme (Continuous Professional Development/Education) and a formal re-accreditation process (e.g., Netherlands, Norway, US). Each doctor must demonstrate continuing education and development and compliance with the requirements of the recertification process (Murgatroyd 2011). Poorly performing physicians identified by these systems are retrained with potential restriction on their licence. These surveillance systems however are complex and not always very effective (Lipner et al. 2013). Accountability and sanctions, while a critical part of the safety armament, are not simple to implement or sustain.

Control of Hazards

The control of known hazards may have more immediate application in primary care. For example, the use of risk control is very important in mental health. It is perhaps unfortunate that the most immediate examples are those which restrain or control people; locked wards, restraint techniques and pharmaceutical control of people are thankfully much more sparingly than in the past. There are however much more subtle methods of risk control that can be applied at a population level and which attempt to control hazards rather than people. For instance the analgesic paracetamol was a common method of suicide and non-fatal self-harm, responsible for many accidental deaths and a frequently cause of hepatotoxicity and liver unit admissions. Legislation introduced by the United Kingdom government in 1998 restricted the pack size to 32 tablets in pharmacies and 16 tablets for non-pharmacy sales. Reducing packet size sounds an implausible approach to reducing the risk of poisoning but many people interviewed after overdoses reported that it was an impulsive act involving the use of available drugs stored in the home. Impulsive acts therefore became less dangerous with smaller packets. Ten years after the changes there had been a significant and sustained reduction in suicide and harm from paracetamol and a similar successful restriction on other paracetomol based products (Box 9.1) (Gunnell et al. 2008; Hawton et al. 2012).

Box 9.1. Withdrawal of Co-proxamol to Reduce Suicide
The extent of fatal poisoning with the analgesic co-proxamol was a concern for many years. The margin between therapeutic and potentially lethal concentrations is relatively narrow. Between 1997 and 1999 co-proxamol was the single drug used most frequently for suicide in England and Wales (766 deaths over the 3 year period). The Committee on Safety of Medicines (CSM) advised that co-proxamol should be withdrawn from use in the UK which took place in December 2007.

A steep reduction in prescribing of co-proxamol occurred in the post-intervention period 2005–7 with the number of prescriptions falling by 59 %. Prescribing of some other analgesics increased significantly during this time. These changes were associated with a major reduction in deaths involving co-proxamol compared with an estimated 295 fewer suicides and 349 fewer deaths including accidental poisonings. During the 6 years following the withdrawal of co-proxamol there was a major reduction in poisoning deaths involving this drug without apparent increase in deaths involving other analgesics.

Adapted from Hawton et al. (2012)

Monitoring, Adaptation and Response

General practitioners and other primary care professionals are of course constantly engaged in monitoring patients but a safety strategy of this nature represents a broader attempt to enhance the capacity to detect deterioration and other problems in the delivery of healthcare to the wider population. As examples we consider one proposal targeted at professionals and a second one aimed at developing a more collaborative culture to help primary care professionals adopt a more patient centred approach and enhance teamwork.

Developing a More Systematic Approach to Watching and Waiting

Time is an essential means of managing risk in primary care. A general practitioner may know from the presenting symptoms that there is a small chance that this is a cancer or other serious complaint; but to refer everyone with such symptoms is neither feasible nor good practice. Instead, they ask the patient to watch and to monitor any change. The use of time is central to the doctor's routines and practice management. Over time diseases and circumstances evolve and a problem encountered at one time will not be the same at a later point. Taking more time solves many health problems in general practice; some will simply resolve, regardless of the diagnosis or intervention, while others will manifest a much clearer symptom pattern. In a significant number of cases, the best way to deal with a situation is just to monitor its development and refrain from clinical intervention. Waiting is paradoxically often more valuable than acting immediately, provided both patient and doctor collaborate in the monitoring of symptoms and trust is maintained on either side. While this strategy is well known and implicitly accepted by both doctors and patients the use of time as a strategy for management has seldom been explicitly studied.

The development of the Tempos Framework (Amalberti and Brami 2012) reflects the importance of time management in primary care. Five time scales termed 'Tempos' requiring parallel processing by GPs are distinguished in the framework: (1) disease's Tempo (unexpected rapid evolutions, slow reaction to treatment); (2) office's Tempo (day-to-day agenda and interruptions); (3) patient's Tempo (time to express symptoms, compliance, emotion); (4) system's Tempo (time for appointments, exams, and feedback) and (5) physician's Tempo (time to access knowledge). This framework (Table 9.1) may serve as a basis for detecting adverse events and recovery, as well as improving adverse event analysis (see Chap. 6).

Improving Transitions Between Hospital and Primary Care

Although the attempt to improve transitions of care has been largely driven from the hospital side it has important implications for safety in primary care. Unintentional changes to medication regimens are an important and well-studied hazard; patients may be discharged from hospital on a very different set of medications from their pre-admission medications, not because of clinical need but through failure to reassess medication at the time of discharge. Medication reconciliation is a process, usually carried out by pharmacists, in which a full assessment is made of the patient's medication before hospital and the new medicines prescribed in hospital,

Table 9.1 Tempo framework for primary care

Disease's tempo	Misleading pathology evolving moving faster or slower than is typical
	Inappropriate therapeutic action, too slow, not efficient. Unfounded reassurance given to the patient on the basis of standard evolution
	Poor explanations/instructions given to the patient and relatives on what should occur, when, what makes an alerting pattern, and what to do.
Doctor's Tempo	Experiencing difficulties in accessing the right knowledge at the right time, due to misleading symptoms, fatigue, pressure or interruptions.
	Technique required for clinical intervention not applied with all usual rigor, due to poor practice, interruptions, fatigue, and more
	Medical case not detected as going beyond doctor's competence
Office's tempo	Excessive busy diaries, time pressure
	Interruptions managements, telephone, patients, secretary, and more
	Incomplete traceability of medical data, rushed medical history, writing style limited to minimum
Patient's tempo	Failure to reveal symptoms, minimizing, or postponing the expression
	Poor doctor-patient relationship, conflicts, specific contexts
System's tempo	Delay in getting appointments for examinations (imagery) or with specialists
	Unexpected approach of emergency department in sending the patient home
	Lost information among careers, lost mail, lost message

Adapted from Amalberti and Brami (2012)

to ensure the patient returns home with the correct medication. Medication reconciliation alone, although important, has not been shown to have clinically significant outcomes, such as reducing subsequent hospitalisation (Kwan et al. 2013). Attention has now turned to a fuller assessment of the entire transition process driven by the high rate of early readmission after discharge. In the United States for instance, nearly one in five Medicare patients are readmitted within 30 days of discharge (Rennke et al. 2013).

Programmes to improve transition have a variety of components and there is as yet little consistency of approach in the various programmes studied. Most have a dedicated discharge team, carry out medication reconciliation, and provide guidance and sometimes training to the patient and family. Some however also extend to a dedicated transition nurse or other professional who has the specific responsibility of monitoring the patient's progress after discharge through telephone calls or visits, coordinating other professionals and responding to any signs of deterioration. Studies give few details of how these programmes are funded and how much training is needed for such people. Strikingly, even the most comprehensive programmes made little if any attempt to engage the main primary care providers (Rennke et al. 2013). This broad approach relies partly on improving the reliability of care within the hospital but, from our point of view, the approach after discharge is one of anticipating, monitoring, adapting and responding to patient need; this is a very different safety strategy than often considered in the primary care context which is largely dominated by attempts to improve adherence to guidelines.

Strategies to improve safety in primary care will require many of the components of these transition programmes. A strong emphasis on patient engagement, coordination and cooperation both within and between teams and above all a mind-set of anticipation, monitoring and caring for the patient beyond their immediate hospital stay or primary care encounter. Primary care providers do of course take this view but current systems, or rather lack of systems, make it very difficult to achieve in practice. New roles in the team could be devoted to this coordination and organisation. New posts of practice facilitators and care managers would enhance the capacity of the primary care team to monitor safety. The care manager's central role could deliver and coordinate services for patients, including coordinating care across clinicians, settings, and conditions, and helping patients access and navigate the system (Taylor et al. 2013).

Mitigation

The capacity to respond rapidly to deterioration is critical to safe care both within hospital and outside, as we have discussed above. The term mitigation extends to the care of patients whose care has failed them in some way leading to harm that has become a new problem in its own right, both for the patient and those caring for them. Dealing with such scenarios of course requires the capacity for rapid response and for all the necessary clinical interventions, but also requires a broader response to deal with the specific problems associated with harm due to poor care rather than to disease.

In other settings we have emphasised the need for support for both patients and staff and this is equally true in primary care. Developing formal programmes to provide such support has been a struggle in hospitals which have the resources and scale to initiate and sustain such help. In primary care, support for either patients or staff relies on the actions of individuals and on responsive and compassionate colleagues. In most cases this is all that is needed but, as we have argued earlier, more extensive and longer term counselling or other interventions may be needed to help patients, careers or staff who have been involved in a serious error or failure. This is currently very difficult to provide in primary care, though some help is given from professional associations. As further integration takes place in primary care we will be able to think more strategically about the management of risk in populations of people and, while prevention and detection will be to the fore, the mitigation of harm should not be neglected.

Reflections on Safety in Primary Care

It may seem premature to think about safety strategies in primary care given the slow pace of development of patient safety in this setting. A review of safety in primary care in the United States published 10 years after the landmark report 'To err is human' found numerous major gaps in understanding of ambulatory safety and almost no credible studies on how to improve primary care safety (Lorincz et al. 2011). There is still a need for basic epidemiological data, for more analyses of the causes of harm to patients in primary care and for the development of specific interventions (Wynia and Classen 2011). We would add that the very concept of patient safety in primary care needs to be examined as, in its current form, it may not resonate sufficiently with primary care practitioners. For patients, safety in primary care is partly associated with control and regulation but strongly linked with personal trust and relationships (Rhodes et al. 2015).

We believe nevertheless that sketching the kind of strategies that might be employed and drawing on a conceptual framework will assist both our understanding of safety in primary care and the development of appropriate intervention strategies. Innovations in information technology will potentially have a massive impact on the coordination of care and the monitoring and support of patients in their homes. Improving systems within practices and clinics and adherence to clinical guidelines are important but may have less impact than in the more structured environment of hospitals. Risk control, in the sense of restricting demand and being clear about competencies and standards, needs to be examined as a formal overall strategy not just in the context of specific clinical issues.

Primary care however is, par excellence, an adaptive system in which clinical decisions evolve from highly individual clinical encounters and relationships in which patient values and preferences are often the pre-eminent consideration. In such contexts we believe that the development of sophisticated monitoring and response strategies may be more important in the overall balance than any of the other broad approaches. The full engagement and indeed education and training of patients and careers in the management of risk will be a necessary core of any such approach. It may be that risk will be more effectively managed in a loose system which incorporates rapid adaptation and response than by the imposition of guidelines and controls. This view however is, as yet, just a view and requires exploration, development and testing.

Key Points
- Primary care faces huge challenges. Primary care practitioners are dealing with increasingly complex conditions making it impossible to provide the best and safest care to every patient.
- Primary care staff can have high personal standards of care without being fully aware of the risks to patients in the wider system of care. This makes it difficult to understand risk at a system level or consider population oriented risk management strategies.
- The concept of patient safety in its current form it may not resonate sufficiently with primary care practitioners. For patients, safety in primary care is strongly linked with personal trust and relationships
- Poor communication and coordination between different elements of the health and social care system, the lack of timely and accurate information after patients are discharged from hospital and delays in obtaining test results are major risks.

- Diagnostic errors have not yet received the attention they deserve, considering their probable importance in leading to harm or sub-standard treatment for patients.
- Quality improvement approaches can be used to improve adherence to guidelines to improve outcomes for patients.
- The exponential growth in scientific knowledge is a particular challenge in primary care. New technologies are expected to assist and support medical decision making and prescribing, ordering and checking test results, and enhance cooperation and coordination
- When systems are under pressure risk control strategies need to be considered to maintain safety and potentially also to constrain costs.
- Time, in the sense of intelligent and active watching and waiting, is an essential means of managing risk in primary care.
- Strategies to improve safety in primary care will require a strong emphasis on patient engagement, coordination between teams and a mind-set of anticipation, monitoring and caring for the patient beyond their immediate primary care encounter.
- Risk in primary care may be more effectively managed in a loose system which incorporates rapid adaptation and response than by the imposition of guidelines and controls.

References

Alderson L, Alderson P, Tan T (2014) Median life span of a cohort of national institute for health and care excellence clinical guidelines was about 60 months. J Clin Epidemiol 67(1):52–55

Amalberti R, Brami J (2012) 'Tempos' management in primary care: a key factor for classifying adverse events, and improving quality and safety. BMJ Qual Saf 21:729–736

Baker R, Hurwitz B (2009) Intentionally harmful violations and patient safety: the example of Harold Shipman. J Royal Soc Med 102:223–227

Barber N (2002) Should we consider non-compliance a medical error? Qual Saf Health Care 11(1):81–84

Bodenheimer T (2006) Primary care: will it survive? New Eng J Med 335(9):861–862

Brami J, Amalberti R (2010) La sécurité du patient en médecine générale. Springer

Buetow S, Kiata L, Liew T, Kenealy T, Dovey S, Elwyn G (2009) Patient error: a preliminary taxonomy. Ann Fam Med 7:223–231

Callen JL, Westbrook JI, Georgiou A, Li J (2012) Failure to follow-up test results for ambulatory patients: a systematic review. J Gen Intern Med 27(10):1334–1348

Croskerry P (2013) From mindless to mindful practice—cognitive bias and clinical decision making. New Eng J Med 368(26):2445–2448

De Lusignan S, Mold F, Sheikh A, Majeed A, Wyatt J, Quinn T, Cavill M et al (2014) Patients' online access to their electronic health records and linked online services: a systematic interpretative review. BMJ Open 4:e006021

Dovey SM, Phillips RL, Green LA, Fryer GE (2003) Types of medical errors commonly reported by family physicians. Am Fam Physician 67(4):697

Gandhi TK, Kachalia A, Thomas EJ, Puopolo AL, Yoon C, Brennan TA, Studdert DM (2006) Missed and delayed diagnoses in the ambulatory setting: a study of closed malpractice claims. Ann Intern Med 145(7):488–496

Gunnell D, Hawton K, Ho D, Evans J, O'Connor S, Potokar J, Donovan J, Kapur N (2008) Hospital admissions for self-harm after discharge from psychiatric inpatient care: cohort study. BMJ 337:a2278

Hawton K, Betgen H, Simkin S, Wells C, Kapur N, Gunnell D (2012) Six-year follow-up of impact of co-proxamol withdrawal in England and Wales on prescribing and deaths: time-series study. PLoS Med 9(5):606

Hogan H, Olsen S, Scobie S, Chapman E, Sachs R, McKee M, Vincent C, Thomson R (2008) What can we learn about patient safety from information sources within an acute hospital: a step on the ladder of integrated risk management? Qual Saf Health Care 17(3):209–215. doi:10.1136/qshc.2006.020008

Hussey PS, Ridgely MS, Rosenthal MB (2011) The PROMETHEUS bundled payment experiment: slow start shows problems in implementing new payment models. Health Aff 30(11):2116–2124

Jacobson L, Elwyn G, Robling M, Jones RT (2003) Error and safety in primary care: no clear boundaries. Fam Pract 20(3):237–241

Jones S, Rudin R, Perry T, Shekelle P (2014) Health information technology: an updated systematic review with a focus on meaningful use. Ann Intern Med 160:48–54

Kahneman D (2011) Thinking, fast and slow. Macmillan, New York

Kret M, Michel P (2013) Esprit, Etude Nationale en Soins primaires sur les événements indésirables, Rapport CCECQA. http://www.ccecqa.asso.fr/projets/esprit

Kripalani S, LeFevre F, Phillips CO, Williams MV, Basaviah P, Baker DW (2007) Deficits in communication and information transfer between hospital-based and primary care physicians: implications for patient safety and continuity of care. JAMA 297(8):831–841

Kwan JL, Lo L, Sampson M, Shojania KG (2013) Medication reconciliation during transitions of care as a patient safety strategy. A systematic review. Ann Intern Med 158(5_Part_2):397–403. doi:10.7326/0003-4819-158-5-201303051-00006

Lee G, Kleinman K, Soumerai S, Tse A, Cole D, Fridkin S, Horan T, Platt R, Gay C, Kassler W, Goldmann D, John Jernigan J, Ashish K, Jha A (2012) Effect of non-payment for preventable infections in U.S. Hospitals. New Engl J Med 367:1428–1437

Lipner RS, Hess BJ, Phillips RL Jr (2013) Specialty board certification in the United States: issues and evidence. J Continu Edu Health Prof 33(Suppl 1):S20–S35

Lorincz CY, Drazen E, Sokol PE (2011) Research in ambulatory patient safety 2000–2010: a 10-year review. American Medical Association, Chicago

Lyratzopoulos G, Wardle J, Rubin G (2014) Rethinking diagnostic delay in cancer: how difficult is the diagnosis? BMJ 349:g7400

Marshall M, Pronovost P, Dixon-Woods M (2013) Promotion of improvement as a science. Lancet 381(9864):419–421

Murgatroyd G (2011) Continuing professional development. General Medical Council, London

Persell S, Dolan N, Friesema E, Thompson J, Kaiser D, Baker D (2010) Frequency of inappropriate medical exceptions to quality measures. Ann Intern Med 152:225–231

Rennke S, Nguyen OK, Shoeb MH, Magan Y, Wachter RM, Ranji SR (2013) Hospital-initiated transitional care interventions as a patient safety strategy. A systematic review. Ann Intern Med 158(5_Part_2):433–440

Rhodes P, Campbell S, Sanders C (2015) Trust, temporality and systems: how do patients understand patient safety in primary care? A qualitative study. Health Expect. doi:10.1111/hex.12342

Ryan AM, Burgess JF, Pesko MF, Borden WB, Dimick JB (2015) The early effects of medicare's mandatory hospital Pay-for-performance program. Health Serv Res 50(1):81–97

Sandars J, Esmail A (2003) The frequency and nature of medical error in primary care: understanding the diversity across studies. Fam Pract 20(3):231–236

Singh H, Giardina T, Meyer A, Forjuoh S, Reis M, Thomas EJ (2013) Types and origins of diagnostic errors in primary care settings. JAMA Inter Med 173:18–25

Shoen C, Osborn R, Squires D, Pasmussen P, Pierson R, Appelbaum S (2012) A survey of primary care doctors in ten countries shows progress in use of health information technology, less in other areas. Health Aff 31(12):2805–2816

Shojania K, Sampson M, Ansari M, JI S, Douvette S, Moher D (2007) How quickly do systematic reviews go out of date? (2007). A survival analysis. Ann Inter Med 147:224–233

Snowdon A, Scnarr K, Alessi C (2014) It's all about me, the personalization of health system. Western University Canada, London

Taylor E, Matcha R, Meyers D, Genevro J, Peikes D (2013) Enhancing the primary care team to provide redesigned care: the roles of practice facilitators and care managers. Ann Inter Med 1:80–83

Vincent C (2010) Patient safety, 2nd edn. Wiley Blackwell, Oxford

Wetzels R, Wolters R, van Weel C, Wensing M (2008) Mix of methods is needed to identify adverse events in general practice: a prospective observational study. BMC Fam Pract 15(9):35

Wachter RM (2010) Why diagnostic errors don't get any respect—and what can be done about them. Health Aff 29(9):1605–1610

Wachter RM (2015) The digital doctor. McGraw Hill, New York

Weiner JP, Fowles JB, Chan KS (2012) New paradigms for measuring clinical performance using electronic health records. International J Qual Health Care 24(3):200–205

Wells KB, Sherbourne C, Schoenbaum M et al (2000) Impact of disseminating quality improvement programs for depression in managed primary care: a randomized controlled trial. JAMA 283(2):212–220. doi:10.1001/jama.283.2.212

Wynia MK, Classen DC (2011) Improving ambulatory patient safety. JAMA 306(22):2504–2505. doi:10.1001/jama.2011.1820

New Challenges for Patient Safety

10

The developments described in the previous chapters are required because our present vision of safety is not adequate for the challenges we face. Our arguments for these developments rest on analyses of the nature of safety in healthcare as it is delivered today. However, as is well known, healthcare is changing rapidly and there are many new opportunities, pressures and challenges. We believe that these coming changes will have further implications for how safety is understood and practiced which will increase the urgency and importance of the transition to a broader vision.

In this chapter we briefly summarise some of the recent and forthcoming developments in healthcare. These have been widely discussed and we are only concerned to summarise some key points. The primary purpose of the chapter is to consider the implications for patient safety and for the strategies and practices we set out in the remainder of the book.

The Changing Nature of Healthcare

The problems faced by healthcare, and many of the challenges for patient safety, arise in part from the very success of modern medicine in combating disease. Because of improvements in diet, nutrition, medicine and environment many people are living longer but also living with one or more chronic conditions such as diabetes, cardiovascular disease and cancer. Diseases which were once fatal are now becoming chronic conditions.

The survival rate for cancers, infections and AIDS, strokes, cardiovascular disease and many other previously fatal diseases have improved significantly even in the last decade. For instance a recent French study of 427,000 new adult cancer cases diagnosed between 1989 and 2004, showed significant improvements in 5 year survival for most cancers, especially prostate cancer (Grosclaude et al. 2013). In the French population of 65 million people over 320,000 new cancers are diagnosed every year; of these 150,000 are designated as 'cured' within the same

© The Author(s) 2016
C. Vincent, R. Amalberti, *Safer Healthcare: Strategies for the Real World*,
DOI 10.1007/978-3-319-25559-0_10

year and a further 150,000 can expect to survive at least 5 years. Similar improvements in survival and quality of life in AIDS patients have been seen in developed countries with the introduction of HAART therapies (Highly Active Antiretroviral Therapy) (Borrell et al. 2006). Most people treated for chronic conditions are going back to work, family and home, with the personal ambition of leading as healthy life as possible. These developments present huge challenges for healthcare systems in providing care and yet remaining affordable.

The traditional hospital cannot remain the main provider of care and core of the medical system simply because it would be unaffordable. Hospitals are still of course essential in any future vision of healthcare but will increasingly focus on investigations and procedures that require a very high level of expertise and sophisticated technology. The proportion of beds devoted to high dependency and intensive care will increase while the overall number of beds will reduce (Ackroyd-Stolarz et al. 2011).

Medical innovations have lead progressively to shorter hospital stays. Earlier diagnosis and less invasive treatments, such as laparoscopic surgery, mean that treatment can be instituted earlier and with less disruption to a person's life. Genomics and preventive medicine will potentially allow even earlier diagnosis and preventative treatment. Increasingly care will need to move outside the hospital which will require a very different vision of primary care. Hospitals specialists will move outside the hospital taking their expertise to homes and to other facilities (Jackson et al. 2013). Because of the growth of point of care testing and the refinement of many treatments, it will be possible to provide a considerable amount of care in community settings. Surgery, radiotherapy, chemotherapy and haemodialysis can all potentially be provided in out-patient settings or smaller community centres.

Box 10.1. A Summary of the Healthcare Paradigm Shift Needed for the Future

From...		...To
One size fits all	Approach	Personalized medicine
Fragmented, One-way	Communication	Integrated, two ways
Provider centred	Focus	Patient centred
Centralized-Hospital	Location	Shift to community
Invasive	Treatment	Less invasive, image-based
Procedure-based	Reimbursement	Episode-based, Outcome-Based
Treating sickness	Objective	Preventing sickness- "Wellness"

Adapted from (http://www.gilcommunity.com/)

The changes outlined above have profound implications for all health professionals (Box 10.1). Over the last 50 years hospital based medical specialties have been dominant in terms of status, reward and expertise. Specialisation has brought the greatest rewards although this has led to a loss of generalist skills and the ability to deal with the complex co-morbidities of care of older patients (Wachter and

Goldman 2002). The need for traditional surgery is declining because of the availability of less invasive interventions carried out by radiologists, gastroenterologists and cardiologists. The role of the doctor is also changing rapidly as more care can be given by nurses and other professionals leaving the doctor in a more supervisory capacity and as the arbiter of complex decisions.

Improved Safety in Some Contexts

While we cannot know exactly what new risks will arise we can at least anticipate some of the areas in which safety may either be enhanced or threatened; some classic hazards will probably decline while others will increase or change in nature. We are mainly concerned with outlining potential new risks but it is important to balance this with an illustration of how innovations and changing patterns of care can bring dramatic improvements in safety.

Healthcare acquired infections have been one of the greatest challenges of recent years and, in some countries, one of the most visible successes in enhancing safety. For instance, surgical site infections are among the most common healthcare associated infections, accounting up to 31 % of healthcare-associated infections in hospitalized patients. However the incidence of clinical significant surgical site infections (CS-SSIs) following low-to moderate-risk ambulatory surgery in low risk patients is declining rapidly through a combination of shorter length of stay and new operative techniques and technologies (Owens et al. 2014). With 80 % of surgery becoming day surgery, nosocomial infection could even become a minor safety issue rather than one that dominates the safety agenda as it has in recent years. This is a radical example of the power of innovation, both in new technologies and organisation of care, in tackling problems that resisted the efforts of even sustained classic quality and safety improvement efforts at the frontline.

Infection and anti-microbial resistance is of course a massive and continuing challenge and remains a major threat to the health of the population, particularly older people with a number of co-morbidities (Yoshikawa 2002; Davies and Davies 2010). We are simply arguing that innovations in surgical care and changing patterns of delivery may well result in a decline in certain types of healthcare acquired infections and therefore a changing pattern of risk.

New Challenges for Patient Safety

Evolution in healthcare, or indeed in any industry, inevitably bring new risks as well as benefits. Some risks arise directly from new technologies and from new forms of organisation. Other risks come, as we have argued, from the very increase in standards that innovation brings as clinical teams and organisations struggle to adapt to the new expectations. For instance, patients are being discharged earlier from hospital after surgery. This is clearly beneficial but, concomitantly, brings new risks. Errors in post-operative care and errors in non-operative management already cause more frequent adverse events than errors in surgical technique (Anderson et al. 2013; Symons et al. 2013). These trends will probably continue and even accelerate.

Increasing Complexity

Evidence based guidelines (mostly developed for people with single diseases) are inappropriate for those people with multiple conditions, resulting in potential over-treatment and over-complex regimes of assessment and surveillance. Problems of harm due to over-treatment and from polypharmacy are likely to increase, exacerbated by the lack of oversight of individual patients in community settings. Clinical judgment becomes more important, not less, as evidence based guidelines become less applicable because of the increasing complexity of patients' illnesses. There is an increased need to listen and determine patients' priorities at the same time as new forms of organisation potentially make this more difficult.

The Challenges and Risks of Care Coordination

The coordination of the care of individual patients, at least those who are more seriously ill, is currently managed through a loose network of hospital doctors, general practitioners and nurses with precise arrangements varying across countries. Care will need to be coordinated and managed much more actively when more is delivered in the community. This will require different models of oversight and a very different organisation of care.

The provision of care to populations of people demands an integration of hospital care, primary care and home care in organisational structures which are already emerging in various forms in England (Dalton 2014). In the United Kingdom general practitioners will struggle to coordinate the increasingly complex care provided. It will be necessary to coordinate high technology resources and services in community clinics to fully supervise patients' health trajectories. Expanded teams and community based care will mean that non-physician providers take on larger responsibilities for patient care.

Patients' pathways are becoming more complex every day. A patient with a chronic condition often has a succession of carers, each for a short period of time, and with a dedicated role. Outside the hospital, and sometimes inside, there may be no overall coordination of care, except through the efforts of the patient and family themselves. Errors resulting from poor coordination between carers and patients are already common (Masotti et al. 2009) and could well increase dramatically. Information technology, team interventions and patient focused solutions can all play a part in the resolution of this issue but the challenge is immense and the solutions difficult to implement.

The Benefits and Risks of Screening

Evidence is mounting that ever earlier detection and ever wider definition of disease is having some adverse consequences for healthy people. Diagnostic scanning of the abdomen, pelvis, chest, head, and neck can reveal "incidental findings" in up to 40 % of individuals being tested for other reasons (Orme et al. 2010). Most of these "incidentalomas" are benign. A very small number of people will benefit from early detection of an incidental malignant tumour, but many others will suffer the anxiety and adverse effects of further investigation and treatment of an "abnormality" that would never have harmed them (Moynihan et al. 2012).

Increased screening also brings more direct hazards. There is evidence from epidemiological studies that the organ doses corresponding to a common CT study (two or three scans, resulting in a dose in the range of 30–90 mSv) result in an increased risk of cancer. The evidence is reasonably convincing for adults and very convincing for children. However 75 % of physicians significantly underestimate the radiation dose from a CT scan, and 53 % of radiologists and 91 % of emergency-room physicians do not believe that CT scans increased the lifetime risk of cancer. It has been estimated that about 0.4 % of all cancers in the United States may be attributable to the radiation from past CT scans. Given the rapid increase in CT scans this estimate might in future be in the range of 1.5–2.0 % (Brenner and Hall 2007).

The Benefits and Risks of Information Technology

The revolution in information technology is having a massive impact on healthcare but also bringing new risks (Wachter 2015). Information technology can reduce risks to patients by providing effective and timely clinical decision support (Jones et al. 2014), improving coordination and communication, and may become a major driver of clinical performance and quality (Weiner et al. 2012; Classen et al. 2011). Various forms of tele-health facilitate and support people in their own homes (Baker et al. 2011; Anker et al. 2011). The massive introduction of IT in healthcare will probably be associated with a reduction of errors due to poor checking, poor readability, and poor traceability (Wachter 2015).

Information technologies are also making decades of stored data usable, searchable, and actionable by the healthcare sector as a whole. This information is in the form of 'big data', so called not only for its sheer volume, but for its complexity, diversity and timeliness. Analysis of big data can help clinicians and organizations deliver higher-quality, more cost-effective care. Big data can potentially lead to the development of an anticipatory health care system, where providers can create personalized evidence-based medicine, tailored to patients' personal preferences for how (Groves et al. 2013).

However such dramatic changes could have negative consequences for both the quality and safety of care if not properly organized, taught to professionals and patients, and accompanied by careful implementation and testing. New risks generated by these technologies are ethics (confidentiality), increased inequalities between regions and social categories, and paradoxically a reduction of direct contact between patients and professionals (Taylor et al. 2014).

Public information on safety will be increasingly available. Public reporting of safety and quality standards is expected to provide accountability and transparency thereby enhancing trust between patients, regulators, payers, and providers (Werner and Asch 2005). Alongside these benefits of public reporting, however, there are potential risks which include a potential loss of trust either in particular institutions or in healthcare more generally. Developing optimal data collection instruments and assuring adequate quality from participating centres are significant challenges (Resnic and Welt 2009). Although considerable efforts are being made to assess safety in a scientific way that allows comparison between hospitals and other facilities, the views expressed on social media could be a much more important determinant of a hospital's reputation.

The Burden of Healthcare: Impact on Patients and Carers

Finally, there is a substantial risk, as care moves into the community, that more demands will be placed on patients and their carers. These demands are potentially quite diffuse and wide ranging as new technologies emerge which are suitable for use in the home. Patients will increasingly have to work collaboratively with hospital and other staff to manage and coordinate their care.

While personal responsibility for care is very important for people who are in reasonable health (Roland and Paddison 2013) it becomes increasingly unrealistic as a person becomes frail and suffering from multiple problems. The burden of organisation of care is greater for patients who are elderly, less well educated, or from less affluent communities or who also have mental health problems. New technology will not solve problems associated with health literacy, which is not likely to improve greatly in the near future. If people are going to be cared for in their homes, both patients and carers will need much more comprehensive support and instructions in the nature of the disease, the treatments they give themselves and most importantly in the detection and response to deterioration.

The phrase 'burden of treatment' refers to the considerable demands that healthcare systems place on patients and carers (Mair and May 2014). For instance patients or their caregivers often have to monitor and manage their symptoms at home, which can include collecting and inputting clinical data. Adhering to complex treatment regimens and coordinating multiple drugs can also contribute to the burden of treatment. Coping with uncoordinated health and social care systems can further add to an ever growing list of management responsibilities and tasks facing patients and their caregivers. This is real work and can be overwhelming—it is time consuming and calls for high levels of numeracy, literacy, and, sometimes, technical knowledge. People who are socially isolated, poorly educated, have low health literacy, are cognitively impaired, do not speak the local language, or who have sensory or physical challenges will simply find this impossible. Mair and May (2014) propose that a key future quality metric will be the extent to which care disrupts people's lives and that a key question for doctors to ask their patients is 'Can you really do what I am asking you to do?'.

A Global Revolution Rather Than a Local Evolution

We can foresee that healthcare systems will change dramatically in the way they are organised and the way care is delivered. We will need different kinds of hospitals with fewer beds, shorter stays, advanced technologies and new competencies. Much more care will be delivered in the home and community, as we cope with extended life expectancy and the rise in chronic conditions.

The consequences for those working in healthcare and the organisation of care are profound. In addition to this people no longer view healthcare as they have in the past and assumptions about what is achievable and what is expected are also changing rapidly. Ageing and well-being are coming to be seen as the right of every citizen with the concomitant expectation of reasonable living conditions, medical support, social rights, pensions, and an ability to maintain a full life in the community. This is an empowering emphasis in most respects but it greatly increases the challenges for healthcare as the demands seem ever growing and sometimes impossible to meet. We are now sometimes seeing a presumption of error and poor care if the outcome does not meet expectations rather than, in the past, an acceptance of the course of the disease with only secondary consideration of the possibility of error.

The patient journey is new for healthcare but already replaced in many people's minds with the concept of a lifetime citizen journey. Medical problems are no longer considered in isolation but in the longer term context of a person's life. Legal aspects of this transformation in mindset are already clearly visible. For instance when a patient is harmed by healthcare and seeks compensations there are legal guidelines for assessing the amount due. This total compensation is assessed on several dimensions which include physical disability, suffering and permanent damage and the impact on personal and professional life, loss of earnings and so on. In France the assessment of compensation used to be restricted, apart from some exceptional cases, to the immediate aftermath of the event with the assumption of recovery in a reasonable time period. However in recent years the legal guidelines on both time period and quality of life have been greatly extended so that compensation can now be made for reduced well-being and quality of life in the mid and long term (Béjui-Hugues 2011)

We will also need in the coming decade to rethink and adapt the surveillance of the healthcare system, learn more from the introduction of electronic information for the purpose of surveillance, develop accreditation methods which encompass patient journeys, assess the impact of the movement of professionals and patients across borders, and last but not least, rethink the whole payment scheme of healthcare to reflect the growing collective and interdependent nature of care delivered to patients. The list might seem long but these are not suppositions about the future but present realities.

These changes, already well underway have important implications for the management of safety in healthcare. We have already argued that we need an expanded vision of safety along the patient journey and which is adapted to multiple contexts. This is already necessary but will be given greater impetus by the changes summarised above and by the inevitable challenges to safety in periods of transition. We believe that we need to try to anticipate the risks both of the new systems and of the transitional period with its inevitable upheavals. The management of risk, and the wider vision of patient safety, needs to be integrated into the new and evolving systems.

Key Points

- The population is ageing due to the advances of modern medicine combined with improved diet and environment. Many people are now living with chronic conditions that were once fatal.
- Multiple innovations in technical care, such as minimally invasive surgery, have significantly shortened hospital length of stay
- Improving standards of care, new technology and new organisations can bring huge benefits but also create new risks and place new burdens on both patients and professionals. Those tendencies are expected to continue and accelerate with the new advent of genomics and personalised medicine.
- A new model of healthcare needs to emerge in which there is a transition from carer and hospital centred rationale to a focus on the patient's journey across settings with much care delivered at home and in the community. These changes are already underway and having a considerable impact on hospitals.
- Some hazards, such as nosocomial infections, should reduce. However we should anticipate new risks such as increased problems in the coordination of care, more problems with over treatment and the integration of multiple treatments in patients suffering from a number of diseases.
- Information technology and personalized medicine are often cited as solutions to these new patient safety problems, but will probably need significant adaptation and maturation before delivering all their potential capacities for safety improvement.
- The 'burden of treatment' may become considerable as more care moves to the home and community. A key question for doctors to ask their patients is "Can you really do what I am asking you to do?"
- We have already argued that we need an expanded vision of safety along the patient journey which is adapted to multiple contexts. This is already necessary but will be given greater impetus by the changes summarised above and by the inevitable challenges to safety in periods of transition.
- The changes required have huge implications for the organisation of healthcare and for the work of professionals. Perhaps most importantly for the healthcare system, it is also a profound change of the whole society, and in the expectations of its citizens.

References

Ackroyd-Stolarz S, Guernsey JR, Mackinnon NJ, Kovacs G (2011) The association between a prolonged stay in the emergency department and adverse events in older patients admitted to hospital: a retrospective cohort study. BMJ Qual Saf 20(7):564–569

Anderson O, Davis R, Hanna GB, Vincent CA (2013) Surgical adverse events: a systematic review. Am J Surg 206(2):253–262

Anker SD, Koehler F, Abraham WT (2011) Telemedicine and remote management of patients with heart failure. Lancet 378(9792):731–739

Baker LC, Johnson SJ, Macaulay D, Birnbaum H (2011) Integrated telehealth and care management program for Medicare beneficiaries with chronic disease linked to savings. Health Aff 30(9):1689–1697

Béjui-Hugues H (2011) La nomenclature Dintihac, de l'évaluation à l'indemnisation. http://www.sante.gouv.fr/IMG/pdf/CNAmed_nomenclature_Dintilhac.pdf

Borrell C, Rodríguez-Sanz M, Pasarín MI, Brugal MT, García-de-Olalla P, Marí-Dell'Olmo M, Caylà J (2006) AIDS mortality before and after the introduction of highly active antiretroviral therapy: does it vary with socioeconomic group in a country with a National Health System? Eur J Public Health 16(6):601–608

Brenner DJ, Hall EJ (2007) Computed tomography—an increasing source of radiation exposure. N Engl J Med 357(22):2277–2284

Classen DC, Resar R, Griffin F, Federico F, Frankel T, Kimmel N, Whittington JC, Frankel A, Seger A, James BC (2011) Global trigger tool shows that adverse events in hospitals may be ten times greater than previously measured. Health Aff 30(4):581–589

Dalton D (2014) Examining new options and opportunities for providers of NHS care. Department of Health, London

Davies J, Davies D (2010) Origins and evolution of antibiotic resistance. Microbiol Mol Biol Rev 74(3):417–433

Gilcommunity. The CEO's 360 Perspective. Healthcare 2020. http://www.gilcommunity.com/files/6313/6251/3856/360_perspective_Healthcare_2020.pdf

Grosclaude P, Remontet L, Belot A, Danzon A, Ramasimanana Cerf N, Bossard N (2013) Survie des personnes atteintes de cancer en France, 1989-2007, Rapport Invs, février 2013

Groves P, Kayyali B, Knott D, Van Kuiken S (2013) The 'big data' revolution in healthcare. McKinsey Q. Available at http://www.mckinsey.com/insights/health_systems_and_services/the_big-data_revolution_in_us_health_care

Jackson GL, Powers BJ, Chatterjee R, Bettger JP, Kemper AR, Hasselblad V, Dolor RJ, Irvine RJ, Heidenfelder BL, Kendrick AS, Gray R, Williams JW (2013) The patient-centred medical home: a systematic review. Ann Intern Med 158(3):169–178

Jones SS, Rudin RS, Perry T, Shekelle PG (2014) Health information technology: an updated systematic review with a focus on meaningful use. Ann Intern Med 160(1):48–54

Mair FS, May CR (2014) Thinking about the burden of treatment. BMC Health Serv Res 14:281

Masotti P, Green M, McColl MA (2009) Adverse events in community care: implications for practice, policy and research. Healthc Q 12(1):69–76

Moynihan R, Doust J, Henry D (2012) Preventing over diagnosis: how to stop harming the healthy. BMJ 344:e3502

Orme NM, Fletcher JG, Siddiki HA, Harmsen WS, O'Byrne MM, Port JD, Tremaine WJ, Pitot HC, McFarland EG, Robinson ME, Koenig BA, King BF, Wolf SM (2010) Incidental findings in imaging research: evaluating incidence, benefit, and burden. Arch Intern Med 170(17):1525–1532

Owens PL, Barrett ML, Raetzman S, Maggard-Gibbons M, Steiner CA (2014) Surgical site infections following ambulatory surgery procedures. JAMA 311(7):709–716

Resnic FS, Welt FG (2009) The public health hazards of risk avoidance associated with public reporting of risk-adjusted outcomes in coronary intervention. J Am Coll Cardiol 53(10):825–830

Roland M, Paddison C (2013) Better management of patients with multimorbidity. BMJ 346:f2510

Symons NR, Almoudaris AM, Nagpal K, Vincent CA, Moorthy K (2013) An observational study of the frequency, severity, and etiology of failures in postoperative care after major elective general surgery. Ann Surg 257(1):1–5

Taylor SP, Ledford R, Palmer V, Abel E (2014) We need to talk: an observational study of the impact of electronic medical record implementation on hospital communication. BMJ Qual Saf 23(7):584–588

Wachter RM (2015) The digital doctor. McGraw Hill, New York

Wachter RM, Goldman L (2002) The hospitalist movement 5 years later. JAMA 287(4):487–494

Weiner JP, Fowles JB, Chan KS (2012) New paradigms for measuring clinical performance using electronic health records. International J Qual Health Care 24(3):200–205

Werner RM, Asch DA (2005) The unintended consequences of publicly reporting quality information. JAMA 293(10):1239–1244

Yoshikawa TT (2002) Antimicrobial resistance and aging: beginning of the end of the antibiotic era? J Am Geriatr Soc 50(s7):226–229

A Compendium of Safety Strategies and Interventions

11

The foundational ideas which have informed our thinking are essentially quite simple but hard won in the sense that they are not for the most part embedded in current thinking. We have argued that we need to view safety through the patient's eyes and that safety needs to be approached very differently in the varying settings along the patient journey. This implies in turn that we need to think more explicitly about what kind of safety strategies are most useful in different contexts. We can now draw these themes together and consider the new directions that emerge.

In this chapter we first review the ideas and arguments of the book and summarise the transitions in patient safety that we believe are needed (Box 11.1). We then set out a compendium of safety and risk management strategies which can be selected, combined and customised to any healthcare setting.

Box 11.1. Five Transitions for Patient Safety
- Understanding risk and harm through the patient's eyes
- Assessing both benefit and harm across episodes of care
- Patient safety as the management of risk over time
- Varying safety models dependent on context
- Using a wider range of safety strategies and interventions

Seeing Safety Through the Patient's Eyes

Our current approach to patient safety, seen from the perspective of healthcare professionals, assumes generally high quality healthcare punctuated by safety incidents and adverse events. This is a sincere vision in that professionals naturally assume that for the most part they are giving good care though they know that there are occasional lapses. In contrast we have endeavoured to see safety through the patient's eyes. A patient may receive wonderful care during one hospital admission, followed by decline due to inadequate monitoring in the community which is later corrected and their health restored; our five levels of care are a formalisation of these varying standards of care that are experienced along the patient journey. This is a vision of safety from the perspective of the patient, carer and family which is the reality we need to capture (Box 11.2).

Box 11.2. Seeing Safety Through the Patient's Eyes
- Isolated errors and incidents are generally less important than the overall coordination of care and the avoidance of major lapses.
- Coordination of care acquires a much greater importance as a safety issue.
- Patients with multiple problems face major challenges in coordinating their own care which can be a considerable burden and source of anxiety.
- Safety interventions to support patients at home will need to focus on organisational interventions such as rapid response to crises and coordination between agencies.
- The healthcare system needs to give more attention to the perspective of patients and families in monitoring and maintaining safety.

Most people understand that all healthcare involves a degree of risk. The level of risk that is accepted must be outweighed by the expected benefits and should be openly expressed. Failures in the healthcare system will always occur to some degree but their consequences can be limited by honesty, transparency, early response and mitigation. We believe, though this could be formally researched, that this is the pragmatic view that most people take of their healthcare. Medicine reduces suffering and improves our lives in many ways but is necessarily limited in what it can achieve. What counts for us as patients is whether healthcare improves our lives overall and whether it lives up to our expectations both technically and in the manner in which the care is provided. The engagement and relationship with clinical staff is important in itself but also affects the overall assessment of whether the care has been beneficial or harmful.

Considering Benefit and Harm Along the Patient Journey

Seeing safety through the patient's eyes has the immediate consequence that we need to view safety along the patient journey. This means that we need to examine episodes of care and consider both risk and harm within an extended timescale. We can still of course examine specific incidents occurring at particular times and this remains a useful exercise. However such an approach will not identify all safety issues and it is not well adapted to either understanding or improving safety in community settings. This longer term approach has consequences for the measurement of harm, for methods of analysis and of course for safety interventions.

The measurement of harm has previously focused on examining the incidence of specific adverse events. There is nothing wrong with such an approach; it provides important baseline information in particular settings which can be used to monitor certain types of harm. However these approaches will need to be extended to assess the balance of benefit and harm over time for any one patient and eventually for populations of patients. Indicators of the reliability and overall quality of care across different healthcare settings might include reduction of repeated hospitalisation, time to response to problem, or the wider impact on work and family (Mountford and Davie 2010). Ideally we need information systems that can track patients over time and provide links between different healthcare settings and forms of treatment. In the longer term we need to develop metrics which can assess the holistic contribution of healthcare to a person's life, in which overall benefits and harm can be assessed and combined. This would truly be a patient centred vision in which the totality of healthcare was assessed not simply the disease specific outcomes. This is not going to be at all easy but is, we believe, the direction that we need to take.

Most of our methods of incident analysis have been restricted to relatively short time periods within a single hospital admission, although the basic concepts have proved robust in other settings such as primary care and mental health. We will have to expand these approaches to examine periods of care rather than a specific incident and its antecedents. We do not yet have fully developed methods to conduct safety analyses over long time periods and so new approaches will need to be developed. Initial analyses have shown that very different considerations emerge such as the critical role of the timing of decisions and actions in the clinical process (Amalberti and Brami 2012). These new forms of analysis will need to encompass a timeframe sufficient to embrace initial assessment, provision of treatment, monitoring the result, and responding to complications while continuing to deliver care. These analyses are likely to place a much greater emphasis on the detection and recovery from problems in the delivery of care.

Patient Safety as the Management of Risk Over Time

We have now arrived at a rather different view of patient safety which includes, but does not conflict with, definitions focused on the reduction of error and harm. The revised aim of patient safety is to maximise the overall balance of benefit and harm to the patient, rather than specifically to reduce errors and incidents. Patient safety becomes the management of risk over time as the patient and family move through the healthcare system. The benefit may be expressed as recovery whenever possible, reduction of suffering or extended survival. This is of course the aim of clinicians everywhere when treating individual patients but we are concerned with how this might be achieved across a system.

The reduction of harm remains important, as does the reduction in errors and incidents, but it is not the dominant perspective. Incidents associated with care will always occur during episodes of care since no human activity can be error free, especially across a system with open access 24 h a day and 7 days a week. Harm may occur because of single safety incidents but more commonly from an accumulation of poor care that impedes recovery, worsens the prognosis or prolongs disability unnecessarily. Patient safety is both the art of minimizing these incidents and managing risk over longer time periods which will require additional skills and methods. We accept in this vision that errors will inevitably occur but that, in a safe system, very few will have any consequences for the patient. This is in essence a clinical vision but at the level of the system as well as the individual patient. Note that this view gives considerable emphasis to the achievements of patients, families and staff in monitoring, negotiating, adapting and recovering from the inevitable hazards and failures along the patient journey.

Adopting a Range of Safety Models

Safety needs to be approached very differently in different environments. We have initially distinguished three classes of safety models that fit different field demands: the adaptive model embracing risks, the high reliability model managing risks, and the ultra-safe model in which risk is controlled or avoided wherever possible. These different responses to risk give rise to different models of safety, each with their own advantages and limitations. The differences between these models lie in the trade-off between the benefits of adaptability and the benefits of standardisation and control.

Healthcare has many different types of activity and clinical settings and so we cannot use one primary model (Box 11.3). We can see parallels and applications of the three models relatively easily in the hospital environment. Radiotherapy, blood products, imaging systems and the management of drugs in pharmacy are all highly regulated, very reliable and operate to industrial standards of precision. Many of these systems rely on a high degree of automation and decision support and the professionals working in these areas are accustomed to working in a highly ordered manner. In other settings, such as obstetrics and elective surgery, risk has to be accepted and managed with coordinated teamwork. High risk surgery, trauma medicine and the treatment of rare and dangerous infections require a more adaptive approach though all benefit from a foundation of standard procedures. We should also bear in mind that much adaptation and resilience in healthcare is unnecessary in that it is employed not from clinical necessity but to compensate for wider system deficiencies (Wears and Vincent 2013).

Box 11.3. Safety Models for Healthcare
- There is no one universal model of safety in healthcare that can apply across every setting. Each model has its own advantages, limitations and challenges for improvement.
- The choice of a safety model will derive from professional consensus, from real world experience, an understanding of safety and judgements as to what is politically feasible in the context in question.

- Imposition of a given safety model that is inappropriate to the context in question may not be effective and may sometimes even degrade safety.
- Each model has similar potential to improve safety in healthcare by a factor of 10, although the maximum attainable safety figures are context dependent and can vary considerably from one model to another.

In healthcare we may find we need a wider array of models than the three we have outlined. It would be a mistake to assume that these three broad approaches are all we need; they are a helpful simplification of a more complex problem. For instance care in the community is unusual in being highly distributed amongst different people and organisations and also only partially reliant on strict standards. Many industries would manage a very distributed system by careful standardisation of core procedures but this may not be possible when, for instance, managing the care of people with severe mental health problems in the community. We are also aware that the industries we have chosen to illustrate the differing approaches to safety are high hazard, high technology and, while those who work in them support each other, they are not simultaneously concerned with delivering compassionate care to vulnerable people. We will probably need a more thoughtful approach to the systemic management of risk in the care of people with learning disabilities for instance, which will retain the broader strategic understanding but achieve the objective of managing risk through personal relationships as much as through formal strategy.

We will also need to consider how we can move between models. When, for instance, does a previously adaptive approach become sufficiently embedded and understood to begin the transition to a high reliability approach? In part this comes about from innovation, familiarisation and the building of expertise within a community. Innovative surgery for instance always begins in a context of risk and challenge. As experience grows in, for instance the management of aortic aneurysm, the surgery still carries risks but these are known, understood and managed rather than endured.

A patient's journey crosses many medical settings and services, in different contexts, and therefore is necessarily exposed to the whole range of safety models. Controlling risk not only requires managing each setting and the transitions between settings, but also being alert to the fact that safety interventions that are effective in one setting may adversely affect safety in other contexts. For instance a cautious and restrictive control of laboratory services aimed at reducing error that is effective in raising standards locally, might adversely affect safety more widely through the reduction in the availability of timely laboratory results.

The external environment is also a critical determinant of which approach to safety can be adopted. An ultra-safe system relies not only on internal procedures, standardisation and automation but also on being able to control the external environment and working conditions. This is achieved by limiting exposure to risk, as when an airline grounds flights in bad weather, and also by controlling working conditions so that there are, for example, strict controls on how many hours civil aviation pilots can fly and how long they must rest before flying again. With enough resource this would be achievable in some areas of healthcare, and indeed some areas are already very safe. However if we cannot control the demand and working conditions, we necessarily have to rely on more adaptive approaches to safety; a different model may be intrinsically safer but simply not feasible in a particular context. While civil aviation is indeed a source of inspiration and learning such a model is only currently applicable in a relatively limited set of circumstances in healthcare. The approach taken to safety in any healthcare setting may ultimately depend in part on what is politically feasible which will vary by discipline, organization and jurisdiction.

Developing a Wider Range of Safety Strategies

The dominant vision of safety improvement is to increase the reliability of basic procedures. These might be the standard procedures in operating theatres, the prevention of venous thromboembolism or procedures to minimise central line or other infections. A number of major interventions have shown that with sufficient will, a sophisticated approach to implementation and the necessary resources, reliability can be markedly improved in a least a set of core processes.

We still have very limited safety strategies for dealing with the day to day realities of healthcare. The dangers to patients when staff are working in difficult conditions are sometimes discussed though generally in terms of the need for more staff which may, of course, be a reasonable request; if more staff were available, or their time was better used, then it might be possible to meet core standards. However in healthcare we will never be able to meet basic standards all the time and in all contexts. We need therefore to relinquish the hope that we will ever be able to do this in all circumstances and pose a different question. How can we ensure that care is safe, even if not ideal, when working conditions are difficult? How, for instance, should one manage an emergency department at times of very high workload or during major emergencies when the care of some less seriously ill patients is inevitably delayed or compromised. What strategies are available to a young nurse of doctor faced with an absurd workload, multiple competing demands and many sick patients? People do adapt and cope of course, but on an individual basis rather than with a considered team based strategy. Developing considered approaches to the management of risk in such situations is a priority for the next phase of patient safety (Box 11.4).

Box 11.4. Developing a Wider Range of Safety Strategies
- We should extend our safety strategies to include risk control, monitoring and adaptation, and mitigation
- We must not be ashamed to propose strategies that aim to manage risk rather than optimise care as long as the final result is beneficial to the patient and robust to context.

- Developing and implementing considered team based responses to difficult working conditions will be safer than relying on ad hoc improvisation
- Healthcare uses a very limited set of safety interventions. The limited progress in patient safety is partly due to the underuse of the available strategies and interventions. It is like driving a car and only using first gear.

We also need to consider how best to customise specific safety interventions. For example reviews of studies of interventions to reduce falls have provided conflicting evidence of effectiveness – some studies showed strong effects, others none. Frances Healey and colleagues argued that the conflict is only apparent and due to the fact that two very different kinds of interventions have been tested; some trials adopted a one size-fits-all implementation of a set bundle of procedures while others, in contrast, developed an individualized approach to each patient with responsive care planning and post-fall review. The standard intervention has been shown in large randomized controlled trials to have little effect; the more personalized approach, which stresses an adaptive response to risk, is proving very much more effective. Healey comments that this 'makes complete sense in the context of falls risk being a complex combination of intrinsic and extrinsic factors and personal attitudes to risk, in an acute environment where physical condition and therefore falls risk factors are rapidly changing' (Healey et al. 2014 and personal communication 2015).

A Compendium of Safety Strategies

We have proposed five broad safety strategies each associated with a family of interventions. We have provided illustrations of how each strategy might be applied in hospital, home and primary care. The reality is no doubt considerably more complicated and needs to be further explored. But even now, with incomplete understanding, we can set out a suite of potential interventions to improve safety and manage risk.

Table 11.1 brings together many of the strategies and interventions described in previous chapters and offers some comments on their applicability, current use and challenges for implementation. The strategies and interventions can operate at different levels and have divided these into frontline, organisation and system levels. This is not a complete account by any means as, for one thing, we have not included patients and families as users of these approaches. However it makes the general point that some interventions are more useful on the frontline while others are more useful at system level. Care bundles for instance are a frontline team intervention, although managers and regulators may encourage and even mandate their use. Risk control approaches can be used within a clinical team in deciding not to start an operation unless all the equipment is available. However, most risk control interventions, such as restricting demand or controlling working conditions, will be at organisation or system level and require considerable authority to implement. To be effective of course they also need the backing of frontline staff.

Table 11.1 A compendium of safety strategies and interventions

A Compendium of Safety Strategies and Interventions

An incomplete taxonomy

Strategy	Interventions	Level of Implementation			Degree of use	Challenges
		Frontline	Organisation	System		
Safety as best practice: aspire to standards	Focal safety programme to reduce specific harms	✓			Used ++	Allocate more time to implementation
	Improve reliability of targeted processes	✓			Underused +	Reduce disparity within settings
	Improve continuous professional education to adopt best practices	✓	✓		Used +	Limited time allocated to education and training
	Develop more sophisticated guidelines for complex patients			✓	Underused	Personalised medicine in progress
Improvement of systems and processes	Staff training, assessment and feedback	✓	✓		Used +	Excessive use of temporary staff
	Standardisation and simplification of key processes	✓	✓	✓	Underused ++	Increasing volume of policies and wasteful processes
	IT to support decision making	✓	✓		Used +	Usability and integration into workflow remain problematic
	Automation of processes	✓	✓		Underused +	Reluctance to adopt
	Improved equipment design	✓		✓	Used +	Manufacturers not sufficiently engaged in safety
	Formalising team roles and responsibilities	✓	✓		Used	Models available but seldom implemented
	Standardisation and enhancement of handover	✓			Used	Models available but seldom implemented
	Improve working conditions : light, noise, physical environment	✓	✓		Used +	Ample margin for progress
	Reduce interruptions and distractions	✓	✓		Underused ++	Not considered as a problem
	Improve organisation and level of staffing		✓	✓	Underused +	Economic constraints and fixed professional roles
	Creation of new roles and posts to improve coordination		✓	✓	Underused +	Economic constraints and fixed professional roles

Optimisation strategies

Risk management strategies

Category	Strategy				Rating	Comment
Risk control	Withdraw services		✓	✓	Underused ++	Political constraints and potential adverse social impact
	Reduce demand		✓	✓	Underused ++	Political constraints and potential adverse social impact
	Place restrictions on services		✓	✓	Underused ++	Political constraints and potential adverse social impact
	Place restrictions on individuals		✓	✓	Underused ++	Response often too late and too punitive
	Place restrictions on conditions of operation	✓	✓	✓	Underused +	Does not conform to healthcare culture
	Prioritisation of care either temporarily or permanently	✓	✓		Underused ++	Politically difficult at local level
Monitoring, adaptation and response	Improve safety culture	✓	✓		Underused +	Often advised but seldom effectively implemented
	Improve detection of deterioration	✓	✓		Underused +	In progress with increasing attention to failure to rescue
	Develop emergency response systems	✓	✓		Used +	Many examples but could be more widely employed
	Develop team cross checking and monitoring	✓	✓		Used +	Models available and huge potential for increased use
	Briefings and anticipation of hazards	✓	✓		Used +	Models available and huge potential for increased use
	Improve organisational response to pressures and threats to safety	✓	✓		Underused +	Needs exploration, study and development
	Negotiate response to regulatory demands	✓	✓	✓	Underused	Adversarial relationship between providers and regulators
Mitigation	Policy of explanation, apology and support for injured patients	✓	✓		Used +	Policies exist but practice lags behind
	Rapid response to physical harm	✓	✓		Used +	Rapid response in hospital but may be slower in community
	Psychological support for patients and families	✓	✓		Underused +	Policies exist but practice lags behind
	Peer to peer support programmes for staff	✓	✓		Underused ++	Models exist but few examples of effective implementation
	Formal support and mentoring for staff	✓	✓		Underused ++	Models exist but few examples of effective implementation
	Insurance of staff and organisation against claims	✓	✓	✓	Used ++	Widely used but not linked effectively to safety initiatives
	Proactive response to complaints and claims	✓	✓		Underused ++	Models exist but few examples of effective implementation
	Proactive response to media	✓	✓	✓	Underused	Some examples of good practice but frequently difficult
	Open dialogue with regulators	✓	✓	✓	Underused ++	Huge scope for improved and more productive relationships

We realise that these proposals are just a starting point in that considerable work is needed to map and articulate the full range of strategies and interventions that are currently in use and which might be adopted. This has been done for 'best practice' approaches, and to some extent for interventions to improve the system. But we need a much fuller description of all types of strategy and intervention if we are to develop a truly comprehensive approach to safety.

We can point to similar developments in other fields which may serve as a model for how this might be done. There is, for instance, enormous interest in influencing the behaviour of people in a variety of ways; these include diet, smoking, exercise, road safety, the payment of taxes and a host of other policy objectives. There are numerous psychological and social theories which purport to explain changes in human behaviour through a variety of mechanisms each with implications for intervention. In weight loss for instance one might seek to enhance self-esteem as a means of increasing adherence to a diet or place more emphasis on extrinsic motivations such as offering financial incentives (Box 11.5). Susan Michie and colleagues have developed the Behaviour Change Wheel (BCW), a synthesis of 19 frameworks of behaviour change found in the research literature (Michie et al. 2013). The BCW has at its core a model of behaviour known as COM-B standing for capability, opportunity, motivation and behaviour. The BCW identifies different intervention options that can be applied to changing each of the components and policies that can be adopted to deliver those intervention options.

Box 11.5. Contrasting Approaches to Changing Risky Behaviour

Suppose one wished to reduce the propensity of young drivers to engage in risky driving practices such as driving too fast. One would canvass all the options including improving their 'capability' to read the road and adjust their driving to the conditions, restricting their 'opportunity' to drive recklessly by means of speed limiters or speed humps, and establishing whether a promising approach would be to try to change their 'motivation' to drive safely through mass media campaigns or legislation and enforcement. Any or all of these may have some effect. The Behaviour Change Wheel provides a systematic way of determining which options are most likely to achieve the change required.

Adapted from Michie et al. (2014)

Changing behaviour is of course one way of managing risk, particularly in respect of adherence to safety critical procedures. However, in this context, we are drawing a broader parallel with the strategic approach to classifying, interpreting and designing interventions. Michie and colleagues point, as we do, to the plethora of potential interventions, to the fact that most interventions are used singly or in limited combinations. Their approach has been to draw out the distinguishing features of each approach, to classify and integrate in a broad conceptual framework of behaviour change interventions.

Our 'incomplete taxonomy' is a first step towards a similar initiative in the systemic management of risk in healthcare and potentially in other settings. We now need to map the landscape, assess the distinctive assumptions and approach of each strategy and intervention and begin to consider how to customise and combine the interventions to the challenges facing us. At the moment, in most cases, we are only using a fraction of the potential interventions open to us. Drawing on the full range and intervening at all levels of the system would give us much more leverage and power in confronting the challenges of keeping healthcare safe in a time of austerity and rising demand.

Key Points

- There are five major transitions between the current vision of patient safety and the broader one we need for the future.
- Our current approach to patient safety assumes generally high quality healthcare punctuated by occasional safety incidents and adverse events; this as a vision of safety from the perspective of healthcare professionals. We need to also understand risk and harm through the patient's eyes
- Viewing safety through the patient's eyes has the immediate consequence that we need to view safety in the context of the patient journey. This means that we need to examine episodes of care and consider both benefit and harm within an extended timescale.
- Patient safety is the art of minimizing incidents but also managing risk over longer time periods which will require additional skills and methods. We accept in this vision that errors will inevitably occur but that, in a safe system, very few will have any consequences for the patient.
- Safety needs to be approached very differently in different environments. Healthcare has many different types of activity and clinical settings and so we cannot use one primary model.
- We need to develop a wider range of safety strategies and interventions. We should extend our safety strategies to include risk control, monitoring and adaptation, and mitigation
- We have very limited safety strategies for dealing with the day to day realities of healthcare. People adapt and cope, but on an individual basis rather than with a considered team based strategy. Developing considered approaches to the management of risk in such situations is a priority for the next phase of patient safety.
- A compendium of safety strategies and interventions is already available. The slow progress in patient safety is in part due to the fact that we are not using the full range of interventions available. It is like driving a car using only first gear.
- Considerable work is needed to map and articulate the full range of strategies and interventions that are currently in use and which might be adopted.

References

Amalberti R, Brami J (2012) 'Tempos' management in primary care: a key factor for classifying adverse events, and improving quality and safety. BMJ Qual Saf 21(9):729–736

Healey F, Lowe D, Darowski A, Windsor J, Treml J, Byrne L, Husk J, Phipps J (2014) Falls prevention in hospitals and mental health units: an extended evaluation of the FallSafe quality improvement project. Age Ageing 43(4):484–491. doi:10.1093/ageing/aft190

Michie S, Richardson M, Johnston M, Abraham C, Francis J, Hardeman W, Eccles MP, Cane J, Wood CE (2013) The behaviour change technique taxonomy (v1) of 93 hierarchically clustered techniques: building an international consensus for the reporting of behaviour change interventions. Ann Behav Med 46(1):81–95

Michie S, Atkins L, West R (2014) The behaviour change wheel a guide to designing interventions. Silverback Publishing, London

Mountford J, Davie C (2010) Toward an outcomes-based health care system: a view from the United Kingdom. JAMA 304(21):2407–2408

Wears R, Vincent CA (2013) Relying on resilience: too much of a good thing? In: Hollnagel E, Braithwaite J, Wears R (eds) Resilient health care. Burlington, VT: Ashgate. pp 135–144

Managing Risk in the Real World

12

We have put forward a series of arguments culminating in the idea that patient safety should be viewed as the management of risk over time. We have suggested that healthcare could draw on a much wider repertoire of strategies and interventions to manage risk and enhance safety. This has been a book of ideas and argument but we hope that these are both rooted in practice and have practical application. In this chapter we first consider some of the more immediate implications as we see them and then consider the form a longer term exploration and development might take.

Implications for Patients, Carers and Families

The engagement of patients in patient safety has been a slow and difficult process. Much of the initial effort has gone into engaging patients alongside staff in reporting and acting on safety issues. This has been a valuable exercise but there is always (rightly) going to be a limit on what it is reasonable or feasible for patients to take on in hospital. We should now turn our attention to the home and community which will pose very different safety challenges. For instance, nosocomial infections are common in hospitals but we have developed effective ways of countering them which rely on close monitoring and a rapid clinical and organisational response. In the home, the risk of nosocomial infections may be less but other risks arise from the open environment, frequent visitors and varying standards of hygiene. Safety is a moving balance between accepted risks and available solutions; you can improve safety either by changing the exposure to risk or improving solutions.

In the home and community patients are in charge of care, and therefore responsible for safety, capable of making errors and being influenced by the many factors that affect safety. This is more than engagement, shared decision making or partnership. Patients and families are taking on roles and responsibilities that are in other settings restricted to professionals. This raises a host of issues for the management of risk and indeed for the delivery of services generally.

© The Author(s) 2016
C. Vincent, R. Amalberti, *Safer Healthcare: Strategies for the Real World*,
DOI 10.1007/978-3-319-25559-0_12

We know that patients and families take safety very seriously and are ingenious in managing many potentially dangerous scenarios. We have given examples in the book and no doubt many more could be collected and studied to reveal novel strategies and interventions which could be shared, adapted and potentially used more widely. Our five strategies can be used to pose some immediate questions about the risks managed by patients and families. What training should be given? If a professional needs training to, for instance, change a dressing while maintaining sterile conditions then surely patients and carers need training too. To what extent can standards of hygiene be relaxed simply because a sick person has moved from hospital to home? We may need to consider setting standards and controlling the environment in which care can be delivered. What kind of support do patients and families need if they are to monitor safety and act appropriately on signs of deterioration? The example of home haemodialysis given earlier shows that advanced units are now including a suite of safety strategies in their training for patients and families. This could potentially be replicated, in varying degrees of intensity, for other forms of care outside hospital.

Implications for Frontline Clinicians and Managers

In healthcare the word frontline is generally taken to mean clinical staff in direct contact with patients and whose actions and decisions have immediate effects. Managers do not deliver treatment and so are not frontline in that sense. They are frontline however in the sense that the actions of clinical managers have a very powerful influence on safety. A bed manager in a large hospital for instance is constantly juggling patients and beds, assessing the latest request for an urgent bed, trying to place patients in wards that are at least reasonably appropriate and preventing very sick patients being in wards where the staff are not familiar with their needs. 'Being in the wrong place' is high risk if you are very sick. Clinical managers have a huge influence on safety but we know little about the strategies they use.

Both clinicians and managers can do a great deal to improve the standards and value of incident analysis. In the United Kingdom at least what was once an exercise in learning, reflection and improvement has, in some settings, sadly deteriorated into a largely bureaucratic exercise producing numerous recommendations that can never be implemented. There is an urgent need to return to the original purpose of incident analysis, focus on the comprehensive investigation of a much smaller number of events and consider the findings in the context of an overall safety and quality improvement programme. This can all be achieved with methods we already have. We also however need to explore the analysis of episodes of care with the attendant attention to contributory factors at different points, adaptation and recovery from problems, and much greater attention to the accounts of patients and families.

We believe that patients and families should select a proportion of the analyses and be encouraged to contribute as much as they can to analyses; their perspective is obviously particularly critical outside hospital. Their perspective will help us understand the longer term safety problems and to develop new techniques and innovations. This perspective might seem utopian and to require huge resources; it would certainly require some careful organisation and the use of technology to bring in some participants. As before though, quality is more important than quantity. A relatively small number of thorough investigations can produce a huge amount of useful information about the vulnerabilities, defences and resilience of the healthcare system.

Frontline teams, with management support, can initiate a much wider and more strategic programme of risk management than is currently the case. We could envisage the development of a decision tree in which different strategies and interventions could be considered sequentially, both separately and in combination, as candidates to enhance safety in any particular setting and in response to identified problems. Improving standards of practice is the most common approach to safety on the frontline and, if achievable, is an obvious and necessary first step. Next there are multiple ways of improving the wider system, though many are not in the control of frontline teams. A critical task is to identify points in the system where inefficient processes and poor reliability are forcing time wasting and potentially dangerous workarounds; the adaptations are of course necessary at the time but wasteful in that they are simply a compensation for other deficiencies rather than a necessary response to problems or crisis. Coping in the short term is admirable and may be in the best interests of that particular patient but the longer term this attitude is detrimental to safety in that it simply prolongs the underlying problems and removes any incentive for change. Risk controls, achieved with professional and management consensus, protect both patients and staff and could bring order and calm to currently chaotic systems. In emergencies of course risk controls can and should be over-ridden.

Monitoring, adaptation and response can be misused but is nevertheless an absolutely critical safety strategy at every level of the system. A great deal has been achieved in team training in anaesthesia, surgery, emergency medicine and other clinical contexts. The skills of monitoring, cross checking and other features of human factors team training are widely taught and such programmes have been shown to improve safety and clinical outcomes. We need to devote much more energy to understanding how people at every level of the system adapt and respond to safety critical issues and develop methods of preparation and training in these skills.

One important direction of travel would be a parallel exploration of how these and other strategies are used by managers, particularly those directly involved in clinical services. Managers constantly adapt and firefight; how much is necessary and how much unnecessary and due to poor systems? Which strategies and interventions are currently used day-to-day and at times of crisis and which would be optimal? We in no way wish to denigrate the skill and dedication of managers who go to extraordinary lengths to maintain safety. Rather, we want to move away from ad hoc improvisation towards explicit and planned interventions, preparation and training in the use of a portfolio of strategies and interventions. A huge amount could be learned from studying the ways managers adapt and cope and by refining this into a more strategic approach. A customised safety training programme for managers, or perhaps pairs of managers and clinicians, would be high on our wish list for the future of safety.

Implications for Executives and Boards

In the United Kingdom and some other countries boards governing healthcare institutions include people from other sectors who bring very different expertise and perspectives. To an engineer, for instance, it can be very difficult to appreciate that what is tolerated in healthcare is very different from what is tolerated in engineering. Incidents brought to the attention of boards are often understood as horrifying and unusual departures from best practice, rather than as the inevitable by-product of the multiple vulnerabilities of an overstretched system. The most critical realisation at board level is the recognition of the extent of poor reliability, difficult working conditions and the corresponding necessity for ad hoc improvisation and cutting corners that is frequently necessary and often actively encouraged. Even clinical members of boards, who know this from daily experience, may struggle to make this explicit. This is a necessary background understanding to any effective action on safety and the inevitable compromises and trade-offs necessary in the delicate and fluctuating balance between finance, safety, quality and patient experience.

An important observation in the implementation of the recently developed framework for the measurement and monitoring of safety has been that the core ideas appear to resonate in different settings and at different levels of the healthcare system. This is valuable in that an organisation could potentially cohere around a core set of safety questions which are meaningful to staff at all levels. We do not know how our framework of strategies and interventions will be received and to what extent they will be applicable at different levels of the system. We are conscious that the language and practice of safety improvement is more akin to frontline practice, while the language of control, assurance and mitigation are more familiar to those at executive, regulation and policy level. It would be enormously valuable if the safety community could find a language and practice that spanned all levels and contexts, and which resonated with patients, frontline staff, executives and regulators alike. We believe that it is achievable and could provide a much needed clarification and integration of safety initiatives.

Boards too can employ a much wider range of strategies and interventions. A strategic combination of approaches and interventions is necessary to achieve optimal safety in the face of financial restrictions and constraints. An expansion of safety strategies may allow them to employ approaches such as risk control which are more familiar and akin to those employed in the management and oversight of finance. Boards often associate improving safety with spending more money, but a judicious combination of strategies and interventions may allow safety interventions to at least be cost neutral overall. One might imagine for instance that controls, restrictions and improved reliability would reduce costs and permit the development of a programme for managers aimed at optimising the simultaneous management of safety, cost, quality and patient experience. This also may sound optimistic but we believe is possible given a sufficiently wide and well thought out safety programme.

At this level of an organisation the integration of strategies and programmes and the explicit trade-off between objectives is a critical skill. An organisational or regional change strategy is generally a combination of individual sub-programmes developed and led by different directors. The individual programmes almost inevitably conflict with each other. For example the ideal plan for reducing the debt at a satisfactory pace is generally detrimental to investments in staff and new technology and ultimately quality and safety.

The development of the final strategy will rest ultimately with the Chief Executive, the board and a small group of senior leaders. They must arbitrate between the individual directors and programmes, create and maintain an overarching vision which encompasses all the objectives of the organisation. There are good and bad ways of achieving these compromises; each director must be willing to adapt his or her particular programme and integrate with other organisational objectives and plans. The Chief Executive and other senior leaders need to be skilled in arbitrating and negotiating with all concerned to achieve a plan which achieves the objectives of the organisation without unduly compromising frontline quality and safety.

Scenarios of this kind are common currency in business school executive programmes but they very seldom include safety issues, at least for healthcare. Developing scenarios in which safety is managed in a realistic and clear sighted way in the face of financial pressures would be a major step forward in the management of risk. A particularly critical issue is the recognition of the early signs of organisational failure, both for those running organisations and for those attempting to monitor them externally such as regulatory agencies and government. Executive courses aimed specifically at the development of strategies which simultaneously address safety, finance and other organisational objectives are being developed and trialled in the oil and gas industries but have not yet been initiated in healthcare.

Implications for Regulatory Agencies and Government

Regulatory agencies face some major new challenges. Until now most regulation has focused on individual healthcare professionals or specific organisations and institutions. Regulation in its various forms now needs to extend to encompass new organisational forms and the complex series of transitions and interfaces along the patient journey. The accreditation of new types of organisation is already in progress in many countries but still requires further development; it is often not clear, for instance, what jurisdiction regulators have over patients living relatively independently in residential care. Traditional approaches to certification, inspection, and the corresponding evaluation criteria may have to be adapted considerably. To move from accreditation of structures and institutions to accrediting patient journeys across primary, secondary and home care is a huge challenge.

A second major challenge is to find a way of regulating a very rapidly evolving system. Regulators in most other industries are blessed with a relatively static environment in which standards can be set and maintained over years or even decades; there is innovation of course but it does not usually lead to a change in core standards, simply a better way of meeting them. In aviation or the nuclear industry major changes may take 10 years from initial proposal to eventual implementation allowing ample time for the development of professional consensus, formal trials and the gradual absorption into the regulatory framework.

In contrast, the rapid pace of innovation in investigations and treatments in healthcare means that the regulator inevitably lags behind innovation. The fast pace of innovation makes developing new standards very challenging; standards can be developed quickly and adapted to a rapidly changing environment but only with a consequent reduction in rigour and testing, since formal evaluation cannot possibly be achieved within the time available. The present system cannot cope with the pace of innovation but it is far from clear how to develop new and more responsive modes of regulation.

Politicians and others at very senior level are under pressure to maintain the fiction that every citizen can have optimal healthcare. In private at least, it is absolutely critical that government and regulators recognise the vulnerabilities of the system and the gap between what is intended and what is actually delivered. The idea of absolute standards is naïve and potentially dangerous especially for struggling organisations. Innovation and the implementation of new and improved standards, all desirable, place huge pressures on both individual organisations and the wider system and create new safety issues. Many regulatory agencies understand this very well but may nevertheless struggle to find an effective response to the issue. The problem of regulation is often conceived as the problem of finding good ways to detect this gap and identify poorly performing organisations. It is essentially a 'best practice' view of safety. However, the deeper problem of regulation is not so much identifying the departure from standards but about how to manage that gap intelligently and humanely. The problem in our terms is one of monitoring, adaptation and response and to develop approaches that are strategic rather than improvised.

Regulatory agencies have developed very comprehensive approaches to inspection and have devoted most of the energies to monitoring compliance with standards. Much less attention has been given to the critical issue of how to respond when standards are not achieved. In many cases the response seems little more than admonition, threats and re-inspection. A basic risk control strategy would mean closing or limiting facilities when an inspection reveals fundamental problems but this threat is usually met with strong local resistance. The healthcare system either needs to overcome these obstacles or take stock and accept that no facilities cannot and develop a more sophisticated response to lapses in standards. We need, just as at other levels of the system, to consider how organisations and regulators might work together in a process of adaptation and ongoing monitoring of the gap between the ideal and the real. Delay in bringing the organisation to the point of compliance with standards, which can take months or even years, can be dangerous but there is seldom any explicit discussion of how to manage safety in the interim. The art of negotiation of realistic timescales for change and compliance needs exploration, research and development.

Future Directions for Research and Practice

This short book and these proposals are a first step. We believe that there are immediate implications but recognise that if the ideas have merit then they need to be debated, developed further and tested in practice by a community of people. The table in Chap. 11 provides, as we expressed it, an incomplete taxonomy. We know that much more work is need to map the full set of strategies and interventions, assess the value of the overall framework, the nature and purpose of the various interventions and their effectiveness in practice. Our experience so far from the small group of people who generously found time to read an earlier draft is that they recognised the need for a broader view of safety, for a breadth of strategic approach and particularly to the need to customise approaches to safety to different settings and along the patient journey.

The next step is broadly ethnographic. We need to observe, identify and collate safety relevant strategies and interventions at all levels of healthcare organisations and the wider system. Ideally these could be compared and matched with approaches taken in other industries. From there we could develop a more robust taxonomy of approaches and begin to assess which might be applicable in different contexts. A considerable amount of research and empirical work is needed to map the full set of strategies and interventions currently in use, who they are used by and in what context. From this point we could envisage empirical testing of different approaches and combinations of interventions, similar to those already developed for best practice and system improvement but employing a wider repertoire of approaches and, most important of all, being tested at every level of the system.

Many ideas and approaches to safety have been advanced; the very term safety has been contested and defined in numerous ways. We have a plethora of concepts and organisational ideals to guide us on the safety journey. Many of these ideas however have remained as ideas and not found a concrete expression or application. Our approach in contrast, abstract as it may seem to some, is resolutely practical in intention. The safety strategies and approaches we describe are all in use but have not been drawn together in a comprehensive architecture which attempts to embrace all healthcare settings. We have found in previous work that a unifying framework can be valuable to those managing safety at all levels of the healthcare system. We hope that our proposals and the attempt to develop an architecture of safety interventions will be useful now and productive for the future.